The Daily Astronomical and Space Quiz Book

Learn Astronomy with Trivia and Questions that Test Your Knowledge of the Universe

Richard J. Bartlett

Cover Images

Top image:

Messier 45, the Pleiades open star cluster in the constellation of Taurus, the Bull. Credit: Anttler (from Wikimedia.)

Bottom images, from left to right:

Messier 51, the Whirlpool Galaxy in the spring constellation of Canes Venatici, the Hunting Dogs. Credit: Adam Block/Mount Lemmon SkyCenter/University of Arizona.

A hypothetical planet and moon orbiting a nearby red dwarf star. Credit: NASA/Harvard-Smithsonian Center for Astrophysics/D. Aguilar.

The Great Red Spot, a storm larger than Earth in the atmosphere of Jupiter. Credit: NASA / SwRI / MSSS / Gerald Eichstädt / Seán Doran.

The northern lights flowing over the Lyngen fjord in 2012 March. Credit: Ximonic (Simo Räsänen)

The Keck Observatory turns toward the stars on the summit of Mauna Kea in Hawaii. Credit: JPL/Caltech

First Edition, January 2018

Contents

Introduction

About this Book

One way or another, this book has spent an unusually long time in development.

Four years ago, when I first decided to write astronomy books, I started to research astronomical events throughout history. My goal was to write a kind of "on this date" book. Unfortunately, some six months later, I began to realize that I simply didn't have enough notable events to cover all 366 days of the year (including leap years).

Consequently, it was abandoned and left to gather the binary equivalent of dust on my computer's hard drive.

Fast forward to a year ago, January 2017. I'd just (belatedly) finished my annual guides and wanted something light, fun and "quick and easy" to write. I should've known that no such thing exists.

I thought about writing a quiz book but with questions for each and every day of the year. Some would be the astronomical equivalent of general knowledge whereas others could be tied to events that happened "on this date." I could finally use all that data I'd gathered.

Unfortunately, not being content to simply compile a daily list of questions, I also wanted to include graphics and ensure the questions weren't repetitive. I decided upon two questions a day, which meant a total of some 732 questions. Pretty soon just compiling the questions started to become time consuming.

In the end, one month of writing turned into two and then three. I've never taken this long on a book before. I'd completed the majority of it when I finally had to abandon it again because I had other, more time-sensitive projects that were now crying out for my attention.

Finally, seven months later, in mid November 2017, I was able to return and finish the book.

It won't take long, I thought. *It's at least 75% done. I might even have it out before Christmas.*

Not so. Here we are, almost mid January 2018, and I'm just now completing it. At least I can say I'm (almost) happy with the result.

I say "almost" because I wanted to include a small(ish) glossary at the back to help those who aren't so familiar with the terms.

I also wanted to include brief information on the astronomers, the space probes and the moons of the solar system... and then quickly realized there was no way to keep it brief (how do you explain black holes in just one sentence?) and hence, I had to let the idea go.

Besides the time involved, it would also mean increasing the size of the book and it's already fairly hefty.

So here's what we have:

There are two questions for every day of the year and a graphic to accompany the questions. The questions are all multiple choice, with four possible answers (A, B, C and D.)

Sometimes the pictures give a clue to the answer and sometimes, well, they might be red herrings instead.

I took a lot of time selecting the images; they're all relevant to the questions or answers and I tried to find pictures that were interesting, unusual or just plain spectacular. Unfortunately, as much as I'd love to publish this book in color, it's prohibitively expensive and so I've had to settle for a black and white paperback instead.

With that in mind, I've produced a free PDF copy that you can download so you can enjoy the pictures for yourselves. You can download it at the link below. Feel free to share this with whomever you choose.

https://tinyurl.com/astquipdf

It's also worth noting that I very carefully made sure there was no bias toward one answer (eg, C) or another. In all, there are 732 questions, which means there's an equal distribution of 183 correct answers for each choice. (i.e, there are 183 A answers, 183 B answers, etc.)

The answers can be found at the back, but they're not listed in a traditional fashion. Instead of listing them in sequential order (ie, the answers to January 1st, then January 2nd, 3rd, 4th etc) I've grouped all the answers for the 1st together, then all the questions for the 2nd, etc.

For example, to find the answers to the questions for January 1st, you'd look at the answers for the 1st and then the month of January.

I've done this for two reasons:

1. It takes up less space, and
2. You're less likely to see the answers to the next day's questions this way.

Some questions might give the answer to other questions (eg one question might ask who discovered the planet Uranus while another might ask in which year he discovered it.)

Where the answers are the moons of planets, I've provided the name of the parent planet in parenthesis next to it.

Lastly, I've tried to avoid clustering questions based around the same subject matter (eg, the planet Saturn) but this hasn't always been possible. Sometimes, honestly, it was deliberate. For example, if a space probe had passed Jupiter on March 5th, you'll probably see more Jupiter-related questions in that part of the book.

I've done everything I can do to avoid duplication of questions and to avoid incorrect answers. If you find any, please let me know by emailing me at astronomywriter@gmail.com

I hope you find this book interesting, educational and – most of all – fun. It took a long time to compile, but despite the setbacks, I can honestly say I enjoyed writing it.

The Author Online

Amazon US: http://tinyurl.com/rjbamazon-us

Amazon UK: http://tinyurl.com/rjbamazon-uk

Facebook: http://tinyurl.com/rjbfacebook

Twitter: http://tinyurl.com/rjbtwitter

AstroNews: http://tinyurl.com/astronewsus

Email: astronomywriter@gmail.com

Clear skies,

Richard J. Bartlett

January 11, 2018

January

January 1st

An artist's impression of Sirius (left) and its white dwarf companion (right). Credit: NASA, ESA and G. Bacon (STScI)

Who discovered the dwarf planet Ceres on this day in 1801?

A. Galileo Galilei
B. William Herschel
C. Edmond Halley
D. Guiseppe Piazzi

How many light years away is Sirius, the brightest star in the night sky?

A. 4.4
B. 8.6
C. 10.5
D. 11.9

January 2nd

The northern lights flowing over the Lyngen fjord in 2012 March. Credit: Ximonic (Simo Räsänen)

Which astronomer mistakenly announced the discovery of the non-existent planet Vulcan in 1860?

A. Urbain Le Verrier
B. William Herschel
C. Charles Messier
D. John Bevis

In the northern hemisphere the proper name for the northern lights is aurora borealis. What's the name in the southern hemisphere?

A. Aurora Australis
B. Aurora Lumina
C. Aurora Meridionali
D. Aurora Terrarum

January 3rd

Pluto's moon Charon. Credit: NASA/Johns Hopkins University Applied Physics/Southwest Research Institute

In what year was the name Charon officially announced as the name for Pluto's largest moon?

A. 1976
B. 1981
C. 1986
D. 1991

What is the third brightest natural object in the sky after the Sun and Moon?

A. Venus
B. Mercury
C. Jupiter
D. Mars

The Spirit *Mars rover investigates its surroundings on the surface of Mars. Credit: NASA/JPL-Caltech*

The Quadrantids meteors reach their peak in the early hours of this morning. From which constellation do they appear to originate?

A. Canes Venatici
B. Virgo
C. Ursa Major
D. Boötes

In which crater did the *Spirit* Mars rover land on the surface of Mars in 2004?

A. Gusev
B. Graff
C. Galle
D. Greeley

January 5th

An artist's impression of the dwarf planet Eris with the Sun in the distant background. Credit: ESO/L. Calçada

The dwarf planet Eris was discovered today in 2005. What was the nickname given to the world by its discovers?

A. Hermione
B. Xena
C. Santa
D. Gabrielle

Which of the following planets can never appear opposite the Sun in the sky and will therefore never be visible at midnight?

A. Mercury
B. Mars
C. Saturn
D. Neptune

January 6th

This view of a crescent Neptune was taken by Voyager 2 *on August 28th, 1989. Credit: NASA/JPL-Caltech/Kevin M. Gill*

Which astronomer mistakenly observed the planet Neptune in 1613, more than two hundred years before its discovery?

A. Giovanni Cassini
B. Johannes Kepler
C. Johannes Hevelius
D. Galileo Galilei

What happens when you use a barlow lens with an eyepiece on a telescope?

A. It filters out light pollution
B. It doubles the magnification of the eyepiece
C. It reduces the glare from the Moon
D. It halves the magnification of the eyepiece

January 7th

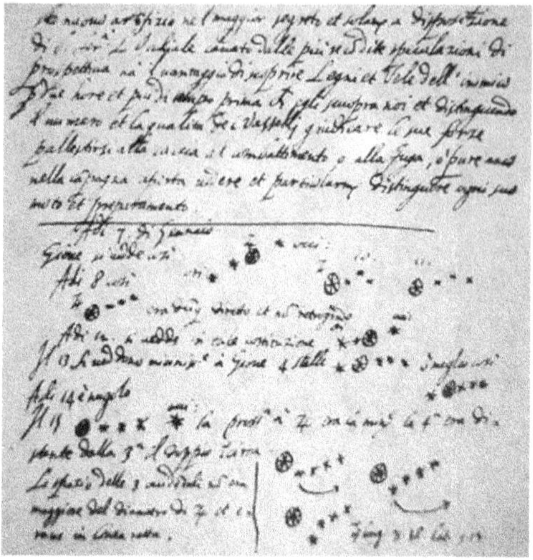

A letter, written by Galileo Galilei, in which he describes his observations of the four largest moons of Jupiter (lower right.) Credit: Public domain

In 1610 Galileo discovered the moons of Jupiter that were later named in his honor. What was his original suggestion for their collective name?

A. Cosmica Sidera
B. Mundus Jovialis
C. Circulatores Jovis
D. Gardes

Which star is *not* a part of the Winter Hexagon asterism?

A. Aldebaran
B. Betelgeuse
C. Capella
D. Sirius

January 8th

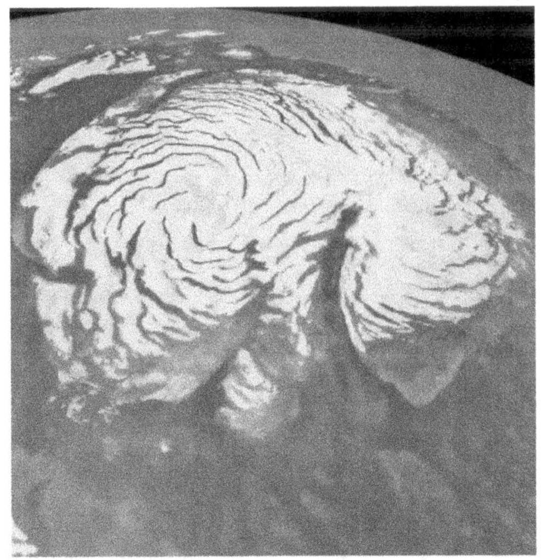

The northern polar ice cap of Mars, imaged by the Mars Global Surveyor. *Credit: NASA/JPL-Caltech/MSSS*

What's the name given to the group of asteroids that precede and follow Jupiter in its orbit?

A. Centaurs
B. Minor Planets
C. Plutoids
D. Trojans

What are the polar ice caps of Mars predominantly composed of?

A. Carbon Dioxide
B. Frozen Oxygen
C. Nitrogen
D. Water

January 9th

Dwarf planet Ceres, imaged by NASA's Dawn *spacecraft. Credit: NASA/JPL-Caltech/UCLA/MPS/DLR/IDA*

Which ancient Greek astronomer theorized that both Mercury and Venus orbited the Sun closer than the Earth?

A. Aristarchus
B. Aristotle
C. Hipparchus
D. Ptolemy

When the dwarf planet Ceres was discovered in 1801, what was it initially thought to be?

A. A new planet
B. A comet
C. A supernova
D. No one knew

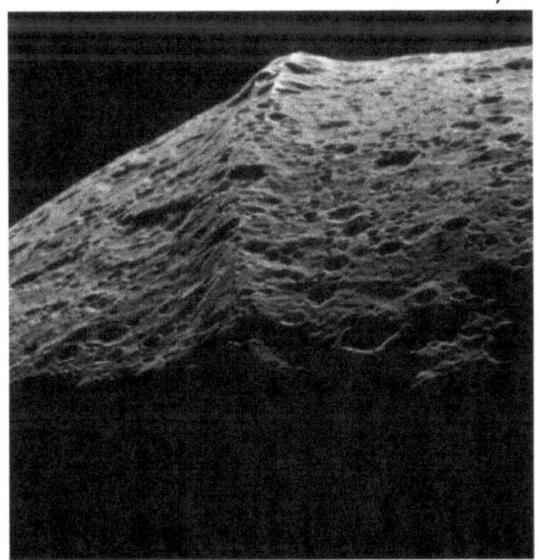

Which moon of Satun has a prominent ridge, as seen in this image by the Cassini *space probe? Credit: NASA*

What's the average distance from Pluto to the Sun? (1 AU = the distance from the Earth to the Sun)

A. 19.2 AU
B. 29.8 AU
C. 39.5 AU
D. 49.0 AU

Which moon of Saturn has a ridge encircling the equator?

A. Dione
B. Iapetus
C. Mimas
D. Titan

January 11th

Uranus's moon Oberon, imaged by the Voyager 2 *space probe. Credit: NASA/JPL*

In what year did William Herschel discover Uranus's moons Titania and Oberon?

A. 1781
B. 1783
C. 1785
D. 1787

Which of the following is *not* a popular name for the Messier 45 open star cluster?

A. The Little Bull
B. The Pleiades
C. The Seven Sisters
D. The Subaru

January 12th

NASA's Deep Impact *probe is launched by a Boeing Delta II rocket. Credit: NASA*

The Great Comet of 1910 was discovered on this day in that year. By what other name was it known?

A. Halley's Comet
B. The Eclipse Comet
C. The Daylight Comet
D. The Southern Comet

In 2005 NASA launched the *Deep Impact* space probe. Which comet was the target?

A. Tempel
B. Holmes
C. Encke
D. Ikeya-Zhang

January 13th

There are two satellite galaxies of the Milky Way that are easily seen with the naked eye from the southern hemisphere. This is the largest. Credit: Gonzalo Vicino

Which of Jupiter's Galilean moons is the smallest?

A. Io
B. Europa
C. Ganymede
D. Callisto

Which explorer has two small satellite galaxies of the Milky Way named after him?

A. Christopher Columbus
B. Captain James Cook
C. Ferdinand Magellan
D. Marco Polo

January 14th

The surface of Titan, as imaged by the Huygens *space probe. Credit: ESA/NASA/JPL/University of Arizona*

In what year did the *Huygens* space probe become the first to successfully touch down on Saturn's moon Titan?

A. 2003
B. 2004
C. 2005
D. 2006

Which comet was the brightest in over 40 years when it appeared in 2007?

A. McNaught
B. Ikeya-Zhang
C. Machholz
D. Hale-Bopp

January 15th

This asteroid was the first to be imaged close-up when Galileo *passed it in 1991. Credit: NASA/JPL/USGS*

In what year did astronomers at Kitt Peak Observatory confirm the discovery of the first pulsar?

A. 1967
B. 1969
C. 1971
D. 1973

What was the first asteroid to be imaged close-up in 1991?

A. Ceres
B. Vesta
C. Ida
D. Gaspra

January 16th

NGC 1987 is a relatively faint open star cluster in the southern constellation of Mensa. Credit: NASA/ESA

In Greek mythology, which of the following constellations was a hunter?

A. Boötes
B. Cepheus
C. Hercules
D. Orion

What does the modern constellation Mensa represent?

A. A Man
B. A Scientist
C. A Table Mountain
D. A Unicorn

January 17th

Despite being only 27 km (16 miles) in diameter, Phobos is the largest of Mars's two moons. Credit: NASA/JPL-Caltech/University of Arizona

In what year was the first precise position of Mars recorded?

A. 272 BC
B. 22 BC
C. 28 AD
D. 253 AD

Roughly how long does Mars's innermost moon, Phobos, have before it is destroyed by tidal forces?

A. 10 – 30 million years
B. 30 – 50 million years
C. 50 – 70 million years
D. 70 – 90 million years

January 18th

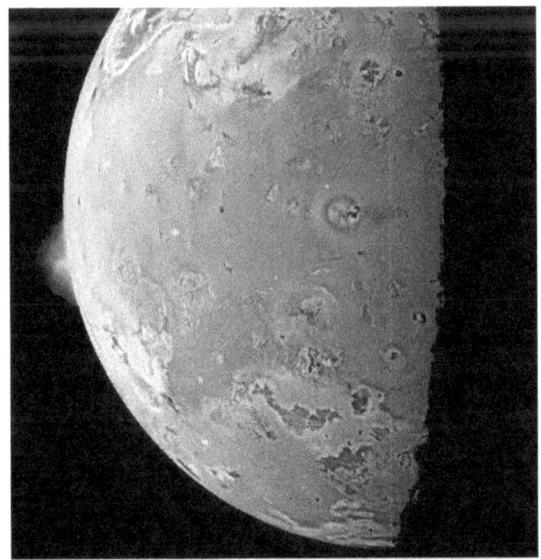

This image from the Galileo *space probe shows a volcanic plume on Io's limb. Credit: NASA/JPL/University of Arizona*

How many comets are known to exist?

A. Between 2,000 and 3,000
B. Between 3,000 and 4,000
C. Between 4,000 and 5,000
D. More than 5,000

Jupiter's moon Io is the most volcanic body in the solar system. How many active volcanoes are on its surface?

A. Less than 300
B. 300 - 399
C. 400 - 499
D. 500 or more

January 19th

The Horsehead nebula is an iconic image in the world of astronomy. Credit: Ken Crawford

In what year did NASA launch the *New Horizons* probe to Pluto?

A. 2003
B. 2004
C. 2005
D. 2006

Which of the following cannot be found in the constellation of Orion?

A. The Bubble Nebula
B. The Horsehead Nebula
C. The Orion Nebula
D. The Running Man Nebula

January 20th

Black holes form when very massive stars collapse at the end of their life. This artist's impression depicts a black hole moving in front of the Large Magellanic Cloud. Credit: Alain R (from Wikimedia)

Which bright star can be found by following the three stars of Orion's belt toward the south?

A. Capella
B. Castor
C. Procyon
D. Sirius

What will the Sun become at the end of its life?

A. A white dwarf
B. A black hole
C. A pulsar
D. A red dwarf

January 21st

The star cluster Messier 41 can be found just below the brilliant star Sirius in winter skies. Credit: Author.

How many objects are commonly accepted to be in Messier's catalog of deep sky objects?

A. 90
B. 100
C. 110
D. 120

What does it mean if a planet is said to be retrograde?

A. It appears to be moving backwards in the sky
B. It appears to be moving forwards in the sky
C. It appears to be opposite the Moon in the sky
D. It appears to be opposite the Sun in the sky

This artist's impression shows the exoplanets orbiting the pulsar PSR B1257+12 - the first exoplanets to be announced. Credit: NASA/JPL-Caltech/R. Hurt (SSC)

In 1992 the first extrasolar planets were announced. In which constellation was the parent star?

A. Hercules
B. Virgo
C. Ursa Major
D. Cancer

In 2014 the European Space Agency announced the detection of water vapor on which dwarf planet?

A. Ceres
B. Eris
C. Pluto
D. Haumea

January 23rd

This sketch of Comet Encke was produced by the renowned British astronomer Rear Admiral William Henry Smyth, circa 1851. Credit: Public domain.

The final transmission from which space probe was received on this date in 2003?

- A. *Pioneer 10*
- B. *Pioneer 11*
- C. *Voyager 1*
- D. *Voyager 2*

Comet Encke has the shortest orbital period of any bright comet. How long does it take to orbit the Sun?

- A. 0.7 years
- B. 1.9 years
- C. 2.6 years
- D. 3.3 years

January 24th

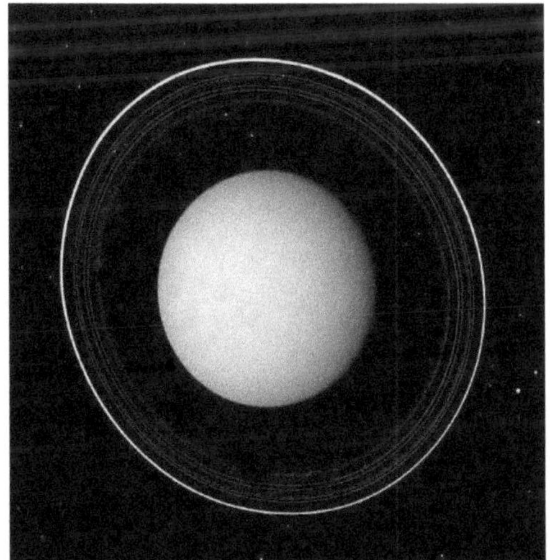

An enhanced image of Uranus, as imaged by the space probe Voyager 2. *Credit: NASA/Erich Karkoschka (U. Arizona)*

In what year did *Voyager 2* make the first successful flyby of the planet Uranus?

A. 1985
B. 1986
C. 1987
D. 1988

Which planet is sometimes called a "failed star"?

A. Jupiter
B. Saturn
C. Uranus
D. Neptune

January 25th

The Orion Nebula can easily be seen with just the naked eye below the three stars of Orion's belt – even from suburban skies. Credit: Rawastrodata (from Wikimedia)

Which Mars rover landed on the red planet on this day in 2004?

A. *Opportunity*
B. *Spirit*
C. *Challenger*
D. *Curiosity*

The constellation Orion represents a hunter. What does the famous Orion Nebula represent?

A. A club
B. A knife
C. A shield
D. A sword

January 26th

The last quarter Moon occurs roughly one week after full Moon. Credit: Author.

The moon Hi'aka was discovered in 2005. Which dwarf planet does it orbit?

A. Makemake
B. Haumea
C. Eris
D. Salacia

When can the last quarter Moon be seen over the southern horizon?

A. At midday
B. At midnight
C. At sunrise
D. At sunset

January 27th

Pluto's hazy atmosphere can be clearly seen in this image from the New Horizons *probe. Credit: NASA/JHUAPL/SwRI*

Roughly how frequently does the planet Mars appear opposite the Sun in the sky and is therefore visible all night?

A. Every year
B. Every 2 years
C. Every 3 years
D. Every 4 years

Which gas is the primary component of Pluto's thin atmosphere?

A. Helium
B. Hydrogen
C. Nitrogen
D. Oxygen

January 28th

The four largest moons of Jupiter – clockwise, from upper left: Io, Europa, Callisto and Ganymede. Credit: NASA/JPL

For whom are the four largest moons of Jupiter named?

A. Daughters of Zeus
B. Lovers of Zeus
C. Servants of Zeus
D. Sisters of Zeus

What's the astronomical symbol for the planet Neptune?

A. An arrow
B. A lightning bolt
C. Ocean waves
D. A trident

January 29th

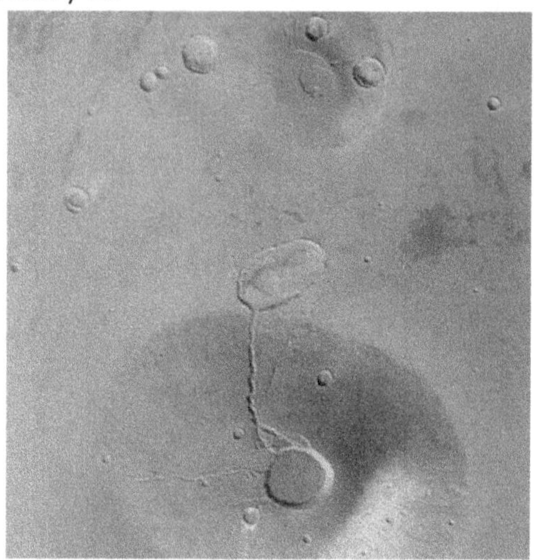

Uranius Tholus (top) and Ceranius Tholus (bottom) are two of the numerous volcanoes to be found on Mars. Credit: NASA/JPL/Malin Space Science Systems

Prior to being renamed in 1831, what was the Royal Astronomical Society known as?

A. The Academy of Astronomy and Science
B. The Astronomical Society of London
C. The Astronomical Society of the United Kingdom
D. The British Astronomical Society

Mars is home to the largest volcano in the solar system. What is its name?

A. Ceraunius Tholus
B. Elysium Mons
C. Olympus Mons
D. Pavonis Mons

The asteroid Eros is the second largest Near Earth Asteroid known to exist. Credit: NASA/JPL/JHUAPL

Roughly how long does Halley's Comet take to complete one orbit of the Sun?

A. 55 years
B. 65 years
C. 75 years
D. 85 years

What percentage of Near Earth Asteroids are thought to be the extinct nuclei of comets?

A. 6%
B. 11%
C. 17%
D. 23%

January 31st

The waxing gibbous Moon. Credit: Author.

Sirius B, the companion to Sirius, is discovered on this day in 1862 by Alvan Graham Clark. What type of star is it?

A. Pulsar
B. Brown Dwarf
C. White Dwarf
D. Red Dwarf

What does it mean if the Moon is said to be waxing?

A. The Moon is due north at sunset
B. The Moon is growing fuller
C. The Moon is becoming thinner
D. The Moon is due south at sunrise

February 1st

Apollo 16 *commander John Young jumps as he salutes the United States flag on the lunar surface. Credit: NASA*

Approximately when is the vernal equinox?

A. December 20th
B. March 20th
C. June 20th
D. September 20th

What is the strength of the Moon's gravity compared to the Earth's?

A. One sixth of Earth's
B. Half of Earth's
C. Three-quarters of Earth's
D. It's the same as the Earth's

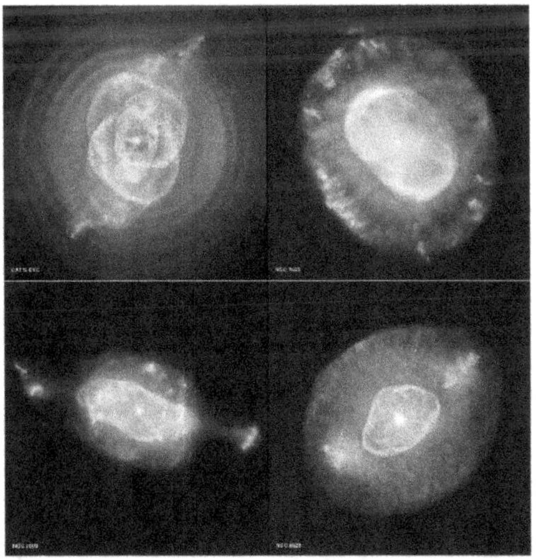

Four examples of planetary nebulae, as imaged by NASA's orbiting Chandra *X-ray observatory. Credit: Smithsonian Institution.*

What is a planetary nebula?

A. A nebula the contains planets
B. A nebula that appears to have a disk like a planet
C. A nebula formed from a dying planet
D. A nebula surrounding a planet

According to the International Astronomical Union, what's the official definition of a fireball?

A. A meteor that's brighter than the Moon
B. A meteor that's brighter than the Sun
C. A meteor that's brighter than Sirius
D. A meteor that's brighter than any planet

February 3rd

This image of the western coastline of North America was taken by the crew of Apollo 16. *Credit: NASA*

What is the rotational period of the Sun at its equator?

A. 25.6 days
B. 30.2 days
C. 35.9 days
D. 40.3 days

What percentage of the Earth's surface is covered by water?

A. 57%
B. 62%
C. 71%
D. 84%

February 4th

The waning crescent Moon. Credit: Author

What did Galileo Galilei discover on this night in 1613?

A. The four largest moons of Jupiter
B. Three of the Trapezium stars in the Orion Nebula
C. The Double Cluster in Perseus
D. The Messier 35 star cluster in Gemini

When is a waning crescent Moon visible in the sky?

A. Afternoon / early evening sky
B. Between sunset and midnight
C. Between mid-evening and midnight
D. Pre-dawn twilight / early morning sky

February 5th

Venus, imaged by the Mariner 10 *space probe using an ultraviolet filter. Credit: NASA*

In what year did the *Mariner 10* space probe send back the first close-up images of the planet Venus?

A. 1972
B. 1973
C. 1974
D. 1975

After Jupiter and Saturn, what is the third largest planet?

A. Earth
B. Mars
C. Neptune
D. Uranus

February 6th

Comet Garradd (lower right) appeared close to the globular star cluster Messier 71 (upper left) in September 2011. Credit: Ravenshoe Group

How many comets are thought to exist in the solar system?

A. One thousand
B. One million
C. One billion
D. One trillion

How many suggested names did Lowell Observatory receive for the yet-to-be named Pluto?

A. More than 1,000
B. Between 500 and 1,000
C. Between 100 and 500
D. Less than 100

February 7th

The rings of Saturn, imaged by the Cassini *space probe. Credit: NASA/JPL*

On this date in 1999 NASA launched the first space probe to return samples of a comet to Earth. What was its name?

A. *Deep Impact*
B. *NEAR Shoemaker*
C. *Rosetta*
D. *Stardust (later NExT)*

How many continuous rings make up Saturn's famous ring system?

A. Six
B. Nine
C. Thirteen
D. Twenty-one

Mars has two small, asteroid-like moons named Phobos and Deimos. Credit: NASA/JPL-Caltech/GSFC/Univ. of Arizona

Which work of fiction predicted the existence of Phobos and Deimos, the two moons of Mars, 150 years before their discovery in 1877?

A. *From the Earth to the Moon*
B. *The Gilded Age: A Tale of Today*
C. *Gulliver's Travels*
D. *The War of the Worlds*

If Uranus were hollow, how many Earths would fit inside?

A. 58
B. 63
C. 68
D. 73

February 9th

Halley's Comet as it appeared during its first return in the 20th century. Credit: Harvard College Observatory

In what year did Halley's Comet make its first of two close approaches to the Sun in the 20th century?

A. 1900
B. 1905
C. 1910
D. 1915

What's the average distance that the Earth orbits the Sun?

A. 135 million km (84 million miles)
B. 140 million km (87 million miles)
C. 145 million km (90 million miles)
D. 150 million km (93 million miles)

February 10th

A waning crescent Moon and Venus rise above the Chilean mountains at Cerro Paranal. Credit: P. Horálek/ESO

What's the maximum distance that Venus will appear from the Sun in our sky?

A. 33.4 degrees
B. 47.8 degrees
C. 56.2 degrees
D. 60.8 degrees

What chemical gives some meteors a blue-green color?

A. Magnesium
B. Iron
C. Sodium
D. Potassium

February 11th

Towards the end of its life the Sun will swell in size and may even swallow the Earth itself. Credit: F. S. Gregs

Roughly how much longer does the Sun have to live?

A. 1 billion years
B. 3 billion years
C. 5 billion years
D. 7 billion years

What's the largest planet in the solar system?

A. Earth
B. Jupiter
C. Saturn
D. Neptune

February 12th

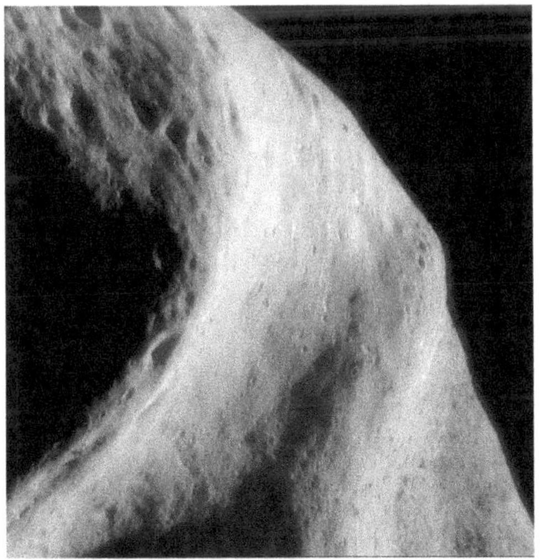

The NEAR Shoemaker *space probe became the first to successfully land on an asteroid. Credit: NASA/JPL/JHUAPL*

On this date in 2001 the *NEAR Shoemaker* space probe becomes the first spacecraft to land on an asteroid. Which asteroid did it land on?

A. Ceres
B. Vesta
C. Mathilde
D. Eros

What is the Chinese translation of Uranus, the name given to the seventh planet from the Sun?

A. Sky King Star
B. Sky Water King
C. Water Ice Star
D. Water King Star

February 13th

A montage of Uranus and its moons, as imaged by the Voyager 2 *space probe. Credit: NASA/JPL*

Sirius is the brightest star in the night sky but which of the following is the *second* brightest?

A. Alpha Centauri
B. Arcturus
C. Canopus
D. Vega

How many moons of Uranus were known before *Voyager 2* began its encounter with the planet in 1985?

A. Three
B. Five
C. Seven
D. Nine

February 14th

The famous Pale Blue Dot photo shows the Earth caught in a beam of sunlight from a distance of roughly six billion km (3.7 billion miles.) Credit: NASA/JPL

In what year was the famous "pale blue dot" photograph of the Earth taken by the *Voyager 1* space probe?

A. 1989
B. 1990
C. 1991
D. 1992

Which of the following constellations represents a flying fish?

A. Fornax
B. Pyxis
C. Volans
D. Vulpecula

February 15th

Mercury (bottom) can be tricky to find but a crescent Moon can be a convenient marker. Credit: Raymond Shobe

When is the best time of year to see Mercury from the northern hemisphere in the early evening sky after sunset?

A. March to early April
B. June to early July
C. September to early October
D. December to early January

Which planet is the closest to the Earth?

A. Jupiter
B. Mars
C. Mercury
D. Venus

February 16th

Messier 97 is one of only four planetary nebulae in the Messier catalog. Credit: Stargazer 7000 (from Wikimedia)

Pierre Mechain discovers the planetary nebula Messier 97 in Ursa Major on this date in 1781. By what name is it better known?

A. Bode's Nebula
B. Coin Nebula
C. Owl Nebula
D. Helix Nebula

If Saturn were hollow, how many Earths could fit inside it?

A. 407
B. 581
C. 635
D. 763

February 17th

This image of asteroid Ida was taken by the Galileo *spacecraft while en-route to Jupiter. Its tiny moon can be seen to the right. Credit: NASA/JPL*

In 1994 asteroid Ida became the first asteroid known to have a moon. What was that moon later named?

A. Aegis
B. Cleoselene
C. Dactyl
D. Romulus

How strong is the lightning on Jupiter compared to the lightning on Earth?

A. 500x stronger
B. 1,000x stronger
C. 1,500x stronger
D. 2,000x stronger

February 18th

Clyde Tombaugh with his homemade telescope, circa 1930. Credit: Public domain.

On this date in 1930, Clyde Tombaugh discovers Pluto on photographic plates taken in January the same year. At which observatory was the discovery made?

A. Kitt Peak
B. Lowell
C. Mount Wilson
D. Palomar

When is Mercury or Venus at superior conjunction?

A. When it is on the opposite side of the Sun
B. When it is between the Earth and the Sun
C. When it is opposite the Sun in the sky
D. When it is in conjunction with an outer planet

February 19th

This image shows the Earth eclipsing the Sun, as seen by the crew of Apollo 12 *on their return home from the Moon. Credit: NASA*

What is an astronomical unit?

A. The distance from the Earth to the Moon
B. The distance from the Earth to Mars
C. The distance from the Earth to the nearest star
D. The distance from the Earth to the Sun

What color is the Sun?

A. Orange
B. Red
C. White
D. Yellow

February 20th

Ptolemy was an ancient Greek astronomer who outlined the original constellations in his book Almagest. *This depiction is from the 16th century. Credit: Public domain.*

On which moon of Jupiter was the first active volcano beyond the Earth discovered?

A. Io
B. Europa
C. Ganymede
D. Callisto

How many of the 88 officially recognized constellations were known to the ancient Greeks?

A. 23
B. 37
C. 48
D. 55

February 21st

An artist's rendition of Olympus Mons surrounded by clouds. Painting by Gordon Legg. Credit: NASA/JPL

In what year was the American Astronomical Society founded?

A. 1884
B. 1889
C. 1894
D. 1899

How tall is the extinct volcano Olympus Mons on Mars compared to Mount Everest on the Earth?

A. Roughly the same height
B. Twice as tall
C. Three times taller
D. Four times the height

February 22nd

An artist's impression of the view from one of the TRAPPIST-1 extrasolar planets. Credit: NASA/JPL-Caltech

The Sun goes through a regular cycle of minimum and maximum sunspot activity. How long does this cycle last?

A. 7 years
B. 9 years
C. 11 years
D. 13 years

On this date in 2017, astronomers announced the discovery of seven exoplanets orbiting a star only 49 light years away. What's special about these planets?

A. At least one shows signs of life
B. They are all Earth-sized and may have water
C. They are all smaller than Earth
D. At least one is known to have a moon

February 23rd

This image from Voyager 1 *immediately drew comparisons to the Death Star from* Star Wars. *Credit: NASA/JPL*

Which moon of Saturn drew comparisons to the Death Star in *Star Wars* when it was first imaged by Voyager 1?

A. Dione
B. Enceladus
C. Mimas
D. Titan

What, or whom, are craters on Mars larger than 60km named for?

A. Ancient Greek astronomers
B. Famous characters from science fiction
C. Gods of war from around the world
D. Scientists and artists whose work involved Mars

February 24th

On a clear, dark night, how many stars might you expect to see? Credit: ESO/H. Dahle

In 2006 astronomers observing the planet Jupiter identify a second dark spot, similar in nature to the planet's famous Great Red Spot. What did they name it?

A. Great Brown Spot
B. Great Dark Spot
C. Pale Red Spot
D. Red Spot Jr.

Under ideal conditions, roughly how many stars might you see from a dark rural location?

A. 1,800
B. 2,700
C. 3,200
D. 4,500

February 25th

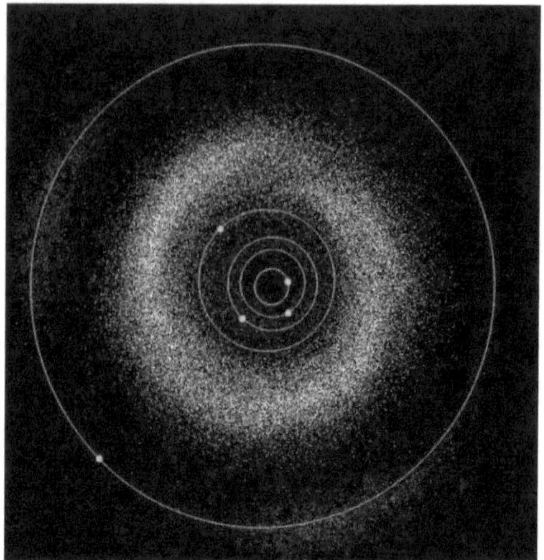

The main asteroid belt lies between the orbits of Mars and Jupiter. Credit: Mdf (from Wikimedia)

What's the name of the international organization that collects data on asteroids and comets?

A. The Asteroid and Cometary Data Center
B. The International Society for Cometary Research
C. The Minor Planet Center
D. The Society for Minor Body Data

How many asteroids are thought to orbit in the main belt between Mars and Jupiter?

A. Less than 500,000
B. Between 500,000 and 750,000
C. Between 750,000 and a million
D. More than a million

February 26th

The Kepler *space observatory has discovered thousands of planets since its launch. Credit: NASA.*

In 2014 astronomers using the *Kepler* Space Telescope announce the largest find of extrasolar planets at that time. How many planets were discovered?

A. 368
B. 468
C. 523
D. 715

Which terrestrial time zone matches Universal Time?

A. Greenwich Mean Time
B. Indian Standard Time
C. Australian Central Standard Time
D. Pacific Standard Time

February 27th

Jupiter and its four largest moons; a montage of images taken by Voyager 1. *Credit: NASA/JPL.*

How long does it take Jupiter to complete one orbit of the Sun?

A. 11.7 years
B. 11.9 years
C. 12.1 years
D. 12.3 years

The names of Jupiter's Galilean satellites – Io, Europa, Ganymede and Callisto – were not widely accepted and in use until which century?

A. 17th Century
B. 18th Century
C. 19th Century
D. 20th Century

February 28th

The Beehive Cluster in the constellation of Cancer is an attractive sight in binoculars and small telescopes. Credit: Eclipse sx (from Wikimedia.)

In 2007, the *New Horizons* spacecraft flew by which planet while en-route to Pluto?

A. Venus
B. Earth
C. Jupiter
D. Saturn

Which open star cluster is also known as the Beehive Cluster?

A. The Butterfly Cluster
B. The Keystone Cluster
C. The Pleiades
D. The Praesepe

February 29th

An artist's impression of a neutron star. Credit: Casey Reed - Penn State University.

On this date in 1504, what celestial event did Christopher Columbus use to intimidate Native Americans to provide him with supplies?

A. A total lunar eclipse
B. A total solar eclipse
C. A meteor shower
D. A supernova

In 1968 the discovery of the first neutron star was announced. Who made the discovery?

A. Fred Hoyle
B. Jocelyn Bell-Burnell
C. Margaret Burbidge
D. Patrick Moore

March

March 1st

The Moon, Mercury (middle) and Venus (upper right) in the twilight skies over Paranal, Chile. Credit: ESO/B. Tafreshi (twanight.org)

Which constellation is the only one based upon a person who once lived?

A. Cassiopeia
B. Cepheus
C. Coma Berenices
D. Hercules

What's the maximum distance that Mercury will appear from the Sun in our sky?

A. 24 degrees
B. 28 degrees
C. 32 degrees
D. 36 degrees

March 2nd

The Philae *lander being deployed at Comet 67P/Churyumov-Gerasimenko. Credit: ESA*

On this date in 2004 the European Space Agency launched the *Philae* lander towards Comet 67P/Churyumov-Gerasimenko. What was the name of the probe that carried it to its destination?

A. *Rosetta*
B. *Stardust*
C. *Deep Impact*
D. *Giotto*

What is a conjunction?

A. When one object passes in front of another
B. When two objects appear opposite one another
C. When three objects appear as a triangle
D. When multiple objects appear close together

March 3rd

This image of Jupiter was taken on December 1st by the first space probe to pass the planet. Credit: NASA.

On this date in 1972, NASA launched its first space probe to Jupiter. Which probe was it?

A. *Pioneer 9*
B. *Pioneer 10*
C. *Pioneer 11*
D. *Voyager 1*

How long does it take Uranus to complete one orbit of the Sun?

A. 82.5 years
B. 83.0 years
C. 83.5 years
D. 84.0 years

March 4th

The nucleus of Comet Halley, imaged on March 4th 1986. Credit: ESA/MPAE, 1986, 1996.

The first images of a comet's nucleus (Halley) were beamed to Earth on this date. Which space probe was responsible?

A. *Giotto*
B. *ICE*
C. *Suisei*
D. *Vega 1*

The Greek word for "atmospheric" gives its name to what type of astronomical phenomenon?

A. Aurorae
B. Eclipses
C. Meteors
D. The Milky Way

March 5th

This spectacular image of a crescent Jupiter was taken by Voyager 1 *as it sped away from the planet. Credit: NASA/JPL.*

In what year did *Voyager 1* make its closes approach to Jupiter?

A. 1978
B. 1979
C. 1980
D. 1981

What precious gemstone might be found deep within the atmospheres of Jupiter, Saturn, Uranus and Neptune?

A. Diamonds
B. Emeralds
C. Rubies
D. Sapphires

March 6th

This artist's impression depicts the first rings discovered about a moon. Credit: NASA/JPL/JHUAPL.

In 2008 scientists announce the discovery of the first tenuous ring to be found about a moon. Which moon was it?

A. Ariel (Uranus)
B. Europa (Jupiter)
C. Rhea (Saturn)
D. Triton (Neptune)

Cassini completed its 100th flyby of which Saturnian moon on this date in 2014?

A. Dione
B. Rhea
C. Mimas
D. Titan

March 7th

The Kepler *orbiting observatory has discovered thousands of exoplanets since its launch in 2009. Credit: NASA/Sandra Joseph, Kevin O'Connell.*

In what year was the dwarf planet Haumea discovered?

A. 2003
B. 2004
C. 2005
D. 2006

The Kepler orbiting observatory was launched on this date in 2009. What is its primary mission?

A. To discover the origins of the universe
B. To discover extra-terrestrial signals
C. To discover the origins of the solar system
D. To discover Earth-sized exoplanets

This montage of Saturn and its moons was created using images from Voyager 1 *and* 2. *Credit: NASA*

In what year on this date did Kepler discover his third law of planetary motion?

A. 1618
B. 1623
C. 1628
D. 1633

What group from Greek mythology are Saturn's moons traditionally named after?

A. The Amazons
B. Norse gods
C. The Titans
D. The Valkyries

March 9th

NASA's Spirit *rover captured this image of a Martian sunset on May 19th, 2005. Credit: NASA/JPL/Texas A&M/Cornell.*

How long is a day on Mars?

A. 22 hours, 53 minutes
B. 23 hours, 29 minutes
C. 24 hours, 40 minutes
D. 25 hours, 17 minutes

What's the average distance from Uranus to the Sun? (1 AU = the average distance from the Earth to the Sun)

A. 19 AU
B. 24 AU
C. 29 AU
D. 34 AU

March 10th

This image of the rings of Uranus was captured by the Voyager 2 *space probe in 1986. Credit: NASA/JPL.*

In what year did astronomers discover the rings of Uranus?

A. 1977
B. 1981
C. 1986
D. 1987

In 2006 NASA announced the discovery of liquid water reservoirs and spouting geysers on which moon?

A. Enceladus (Saturn)
B. Europa (Jupiter)
C. Io (Jupiter)
D. Triton (Neptune)

March 11th

This artist's depiction of the surface of Pluto shows its largest moon, Charon, as a crescent and a dim, noonday Sun. Credit: NASA/Southwest Research Institute/Alex Parker.

How long does it take sunlight to reach Pluto?

A. 3 hours
B. 5½ hours
C. 8 hours
D. 10½ hours

What does it mean if a planet is at opposition?

A. It rises at midnight
B. It sets at sunrise
C. It is opposite the Sun in the sky
D. It is behind the Sun in the sky

The orbits of the known potentially hazardous asteroids that may pose a threat to the Earth (just to the upper right of the Sun). Credit: Kevin Gill.

On a star chart, what's the astronomical equivalent of longitude called?

A. Right Ascension
B. Azimuth
C. Declination
D. Altitude

How many potentially hazardous asteroids are thought to exist?

A. Less than 1,500
B. Between 1,500 and 3,000
C. Between 3,000 and 4,500
D. More than 4,500

March 13th

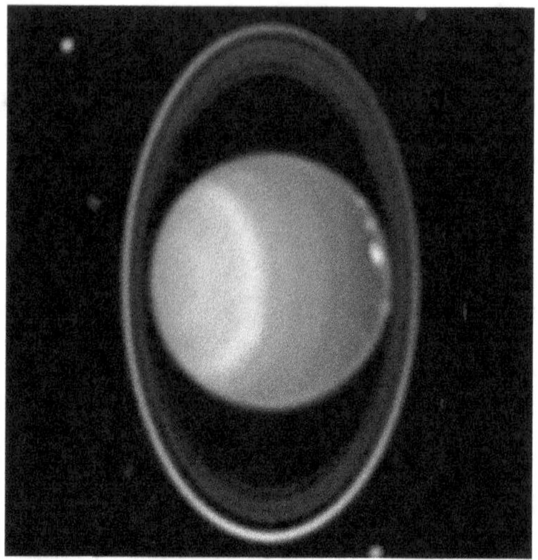

Uranus, as imaged by the Hubble Space Telescope *with the near infrared camera. Credit: Erich Karkoschka (University of Arizona) and NASA/ESA.*

In what year did William Herschel discover the planet Uranus on this date?

A. 1776
B. 1781
C. 1786
D. 1791

Which state designated this day to be Pluto Planet Day in 2007?

A. Arizona
B. California
C. Illinois
D. New Mexico

March 14th

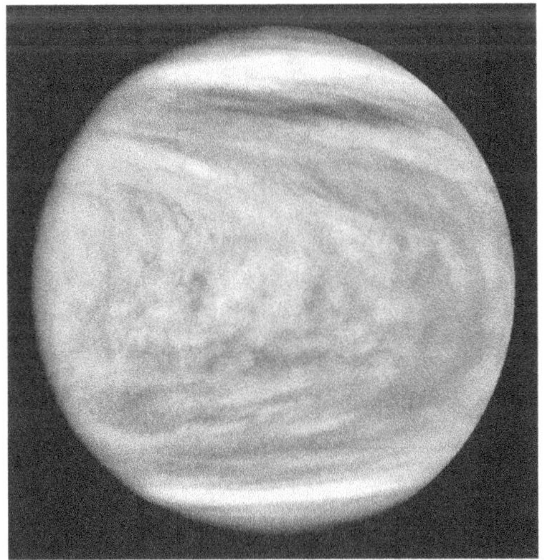

Venus, imaged by the Pioneer Venus *space probe in 1988. Credit: NASA.*

In 1930, upon hearing of the discovery of a new planet, 11 year old Venetia Burney suggested the name Pluto. Her great-uncle had previously named which two moons?

A. Amalthea and Himalia (Jupiter)
B. Enceladus and Mimas (Saturn)
C. Miranda and Titania (Uranus)
D. Phobos and Deimos (Mars)

How long is a day on Venus?

A. 24 hours
B. 243 hours
C. 24 days
D. 243 days

March 15th

Mars was the Roman god of war but his Greek equivalent went by a different name. Credit: Mbdortmund (from Wikimedia.)

In 2004, the discovery of which world prompted the media to report that a tenth planet had been discovered?

A. Eris
B. Haumea
C. Makemake
D. Sedna

Mars was named after a Roman god but who was his Greek equivalent?

A. Ares
B. Honos
C. Osiris
D. Polemos

March 16th

A fireball streaks across the skies over Chile. Credit: ESO/C. Malin.

In what year did *Mariner 10* make its third and final flyby of the planet Mercury?

A. 1973
B. 1974
C. 1975
D. 1976

At what time of year are you 10-30% more likely to see a fireball?

A. Mid March to early April
B. Late June to early July
C. Early to mid August
D. Mid to late December

March 17th

This image from Cassini *shows Phoebe, one of Saturn's smallest moons. Credit: NASA/JPL/Space Science Institute.*

On this date in 1899, which moon became the first to be discovered using photographs rather than visually through a telescope?

A. Phoebe (Saturn)
B. Triton (Neptune)
C. Almalthea (Jupiter)
D. Titania (Uranus)

The visibility of which object can be used to predict rain?

A. The planet Jupiter
B. The Andromeda Galaxy
C. The star Vega
D. The Praesepe, or Beehive star cluster

Mercury is the innermost and smallest planet in the solar system. Credit: NASA/Johns Hopkins University Applied Physics Laboratory/Carnegie Institution of Washington.

Which space probe became the first to orbit Mercury in 2011?

A. Galileo
B. Magellan
C. Mariner 10
D. MESSENGER

Which star is sometimes associated with Electra, the "lost member" of the Pleiades open star cluster?

A. Alcor
B. Aldebaran
C. Altair
D. Antares

March 19th

Jupiter's Great Red Spot is large enough to swallow the entire Earth. Credit: NASA/JPL.

When William Herschel made his first observations of the planet Uranus in 1781, what did he initially believe he'd discovered?

A. An asteroid
B. A comet
C. Another moon
D. A supernova

Wind speeds along the edge of Jupiter's Great Red Spot peak at roughly what speed?

A. 245 kph (152 mph)
B. 317 kph (197 mph)
C. 431 kph (268 mph)
D. 486 kph (302 mph)

March 20th

The Bayeux Tapestry depicted Halley's Comet appearing in the skies prior to the Battle of Hastings. Credit: Myrabella (from Wikimedia).

In what year was Comet Halley observed and seen as a bad omen, prior to the Battle of Hastings and the Norman invasion of England?

A. 1066
B. 1076
C. 1086
D. 1096

What does the word equinox mean?

A. Shortest days
B. Longest nights
C. Equal days
D. Equal nights

March 21st

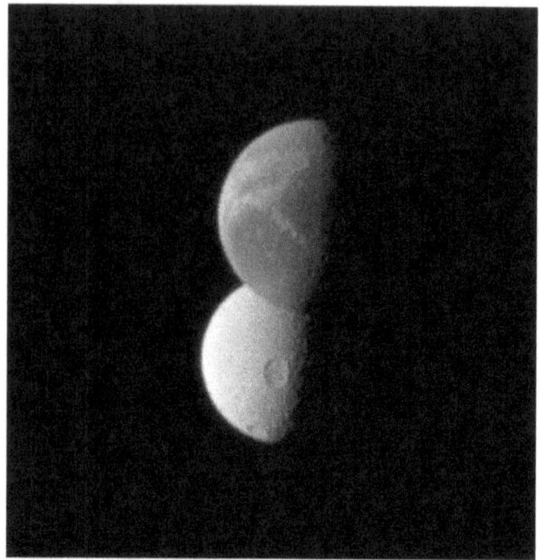

This image from Cassini *shows Dione passing in front of Tethys. Credit: NASA/JPL/Space Science Institute*

Who discovered the Saturnian moons Tethys and Dione on this date in 1684?

A. Giovanni Cassini
B. Galileo Galilei
C. Christiaan Huygens
D. Isaac Newton

What is the latin name for the Sun?

A. Apollo
B. Helios
C. Ra
D. Solis

March 22nd

Comet Hale-Bopp as it appeared over the skies of Croatia. Credit: Philipp Salzgeber.

Which comet made its closest approach to the Earth on this date in 1997?

A. Hale-Bopp
B. Hyakutake
C. Ikeya-Zhang
D. Swift-Tuttle

Mars's moon Deimos is not visible from which area of the planet?

A. The equator
B. The northern hemisphere
C. The poles
D. The southern hemisphere

March 23rd

Meteors have been observed for millennia. This image depicts the Andomedid meteors of November 1872. Credit: Public Domain.

On this date in 687 B.C. Chinese astronomers recorded the first known observation of a meteor shower. Which shower did they observe?

A. Gamma Normids
B. Lyrids
C. Quadrantids
D. Virginids

Which of the following is not a spring constellation?

A. Boötes
B. Canes Venatici
C. Cygnus
D. Virgo

March 24th

It's thought that Stonehenge may have important links to the Sun, especially during the summer and winter solstices. Credit: Simon Wakefield.

In 1993, which comet was the first to be discovered orbiting another planet?

A. Austin
B. Hale-Bopp
C. Shoemaker-Levy 9
D. Swift-Tuttle

Roughly what distance across the sky does the Sun appear to move over the course of 24 hours?

A. 1 degree
B. 2 degrees
C. 5 degrees
D. 10 degrees

March 25th

Apollo 17 *astronaut Harrison Schmitt collects rock samples from the lunar surface. Credit: NASA.*

Which Saturnian moon did Christiaan Huygens discover on this day in 1655?

A. Dione
B. Mimas
C. Rhea
D. Titan

In total, how much lunar rock was returned to Earth by astronauts on the Apollo missions?

A. Less than 200 pounds
B. Between 200 and 500 pounds
C. Between 500 and 800 pounds
D. More than 800 pounds

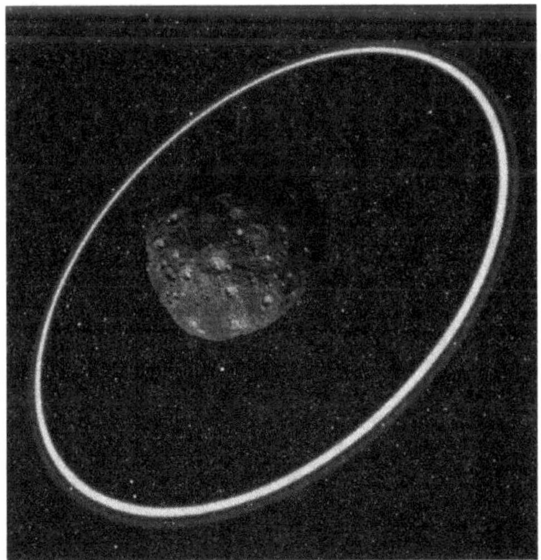

An artist's impression of the first asteroid discovered to have rings. Credit: ESO/L. Calçada/M. Kornmesser/Nick Risinger (skysurvey.org).

In 2014, scientists announce the discovery of the first known ring system about which asteroid?

A. Ceres
B. Chariklo
C. Dionysus
D. Eugenia

Which of the following constellations represents a compass?

A. Circinus
B. Crux
C. Pyxis
D. Volans

March 27th

Arcturus, the brightest star in the constellation of Bootes, the Herdsman, is a prominent sight in the spring night skies. Credit: Space Telescope Science Institute / NASA

Which type of objects are often compared to "dirty snowballs"?

A. Comets
B. Asteroids
C. Dwarf planets
D. Meteors

What's the easiest way of finding Arcturus in the night sky?

A. By looking overhead at midnight, January 1st
B. By following the Milky Way north to Polaris
C. By following the curved tail of Ursa Major south
D. By following the three stars of Orion's belt north

March 28th

Just as the planets orbit the Sun, so the Sun orbits the center of our galaxy. Credit: NASA/JPL-Caltech

In 1802, Heinrich Willhelm Olbers discovered the second asteroid. What was it later called?

A. Ceres
B. Juno
C. Pallas
D. Vesta

Roughly how long does it take the Sun to orbit the center of our galaxy?

A. 200 – 225 million years
B. 225 – 250 million years
C. 250 – 275 million years
D. 275 – 300 million years

March 29th

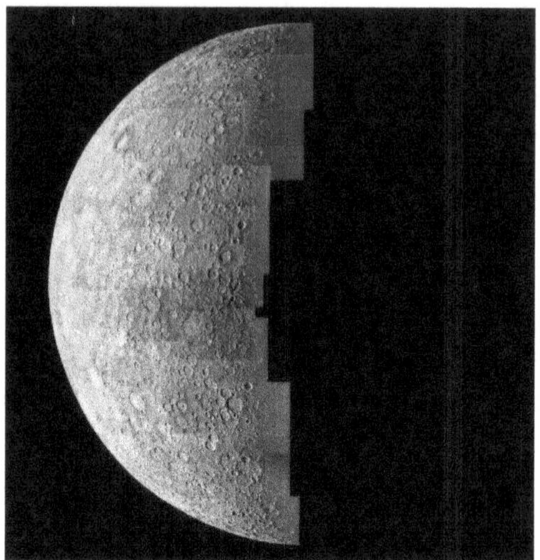

This mosaic of Mercury was created from images sent back from the first space probe to visit the planet. Credit: NASA/JPL

Which space probe was the first to fly by Mercury on this date in 1974?

A. *Mariner 10*
B. *MESSENGER*
C. *Pioneer 9*
D. *Voyager 1*

Who was Giovanni Cassini honoring when he collectively named four of Saturn's moons *Sidera Lodoicea*?

A. Count Panzutti
B. King George III
C. King Louis XIV
D. Pope Leo XI

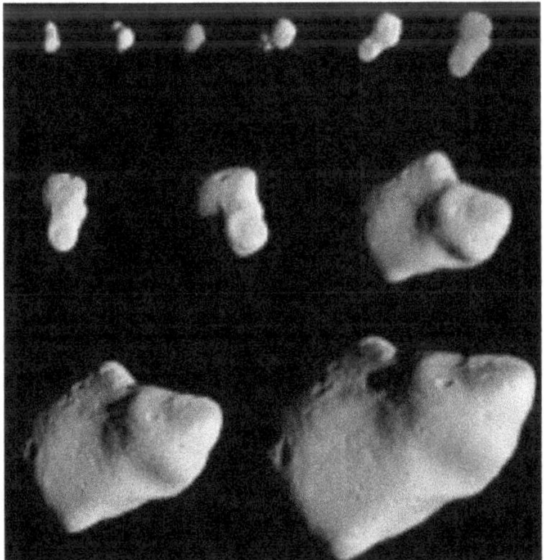

These images of the asteroid Gaspra were taken by the space probe Galileo *in 1991, while en-route to Jupiter. Credit: NASA/JPL.*

Which of the following is another name for asteroids?

A. Dwarf planets
B. Minor planets
C. Planetesimals
D. Plutoids

What's the only planet where you can see the sun rise in the east, then set in the east, rise again and then set in the west?

A. Mercury
B. Jupiter
C. Mars
D. Venus

March 31st

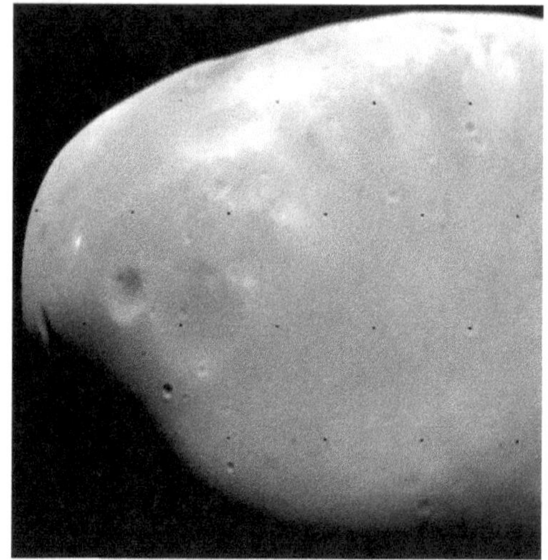

Mars's moon Deimos, imaged by the Viking 2 *orbiter in 1977. Credit NASA/JPL.*

Which dwarf planet was discovered on this date in 2005?

A. Eris
B. Haumea
C. Makemake
D. Sedna

How long does Deimos take to orbit Mars?

A. 30 hours
B. 35 hours
C. 40 hours
D. 45 hours

April

April 1st

The famous "Pillars of Creation" image was taken on this date by the Hubble Space Telescope. *Credit: NASA.*

What unconventional theory did the Soviet astronomer Iosif Shklovksy suggest in 1959?

A. That primitive life existed on Mars
B. That human life had originated on Mars
C. That Mars would appear as large as our Moon
D. That the moons of Mars were artificial

On this day in 1995 the *Hubble* Space Telescope took the famous "Pillars of Creation" photograph. The image features a section of which nebula?

A. Eagle Nebula
B. Lagoon Nebula
C. Orion Nebula
D. Trifid Nebula

The Whirlpool Galaxy, a classic spiral galaxy. Credit: Adam Block/Mount Lemmon SkyCenter/University of Arizona.

Who first observed and described the spiral shape of the Whirlpool Galaxy?

A. Galileo Galilei
B. The third Earl of Rosse
C. William Herschel
D. Charles Messier

How large does the moon Deimos appear in the Martian sky?

A. A bright, starlike point
B. About half the size of our full Moon
C. About size of our full Moon
D. About twice the size of our full Moon

April 3rd

The waxing crescent Moon is a familiar and welcome sight to many amateur astronomers. Credit: Author.

In 2014 scientists announce that one planetary moon may have a subsurface ocean hospitable for life. Which moon was it?

A. Charon (Pluto)
B. Europa (Jupiter)
C. Triton (Neptune)
D. Enceladus (Saturn)

When is a waxing crescent Moon visible in the sky?

A. Evening Twilight
B. Only between 10pm and 2am
C. Only between 2am and 6am
D. Morning Twilight

April 4th

NGC 2997 is a face-on spiral galaxy located approximately 30 million light years away in the constellation Antlia. Credit: ESO/P. Grosbøl.

In 1960, a group of astronomers at the Green Bank Radio Astronomy Observatory begin Project Ozma. What was the project searching for?

A. Black holes
B. Extrasolar planets
C. Signals from extra-terrestrials
D. Pulsars

What does the modern constellation Antlia represent?

A. An Air Pump
B. A Compass
C. An Antelope
D. A Telescope

April 5th

M66 lies about 35 million light years away and is one of the Leo Triplet set of galaxies. Credit: NASA, ESA and the Hubble Heritage (STScI/AURA)-ESA/Hubble Collaboration.

How long is a day on Neptune?

A. 9 hours, 42 minutes
B. 12 hours, 17 minutes
C. 16 hours, 6 mins
D. 23 hours, 56 minutes

NGC 3628 and M66 make up two of the Leo Triplet set of galaxies. What galaxy completes the trio?

A. M31
B. M49
C. M51
D. M65

Mercury, despite being the smallest, is the densest of all the planets. Credit: NASA-APL

On this date in 1973, NASA launches the first space probe to visit both Jupiter and Saturn. Which probe was launched?

A. *Pioneer 10*
B. *Pioneer 11*
C. *Voyager 1*
D. *Voyager 2*

Which element is thought to be the primary component of Mercury's core?

A. Carbon
B. Iron
C. Mercury
D. Nickel

April 7th

Saturn's moon Dione against the golden clouds of the ringed planet. Credit: NASA/JPL/Space Science Institute.

On this date in 2010, the space probe *Cassini* detects a thin exosphere surrounding which Saturnian moon?

A. Dione
B. Enceladus
C. Tethys
D. Titan

Which star's name means "prince" or "little king"?

A. Arcturus
B. Mizar
C. Regulus
D. Spica

April 8th

M101, the Pinwheel Galaxy, is about twice the size of our own Milky Way. Credit: European Space Agency & NASA.

Which galaxy, originally discovered by Charles Messier in 1781, was "lost" for three years and then recovered by William Herschel on this date in 1784?

A. M101
B. M91
C. M61
D. M81

What name did Giovanni Cassini give to the four moons of Saturn he discovered between 1671 and 1684?

A. Cor Caroli
B. Cosmica Sidera
C. Sidereus Nuncius
D. Sidera Lodoicea

April 9th

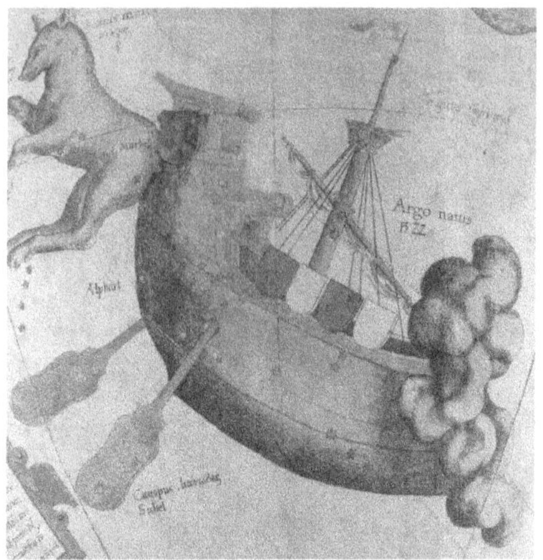

The now defunct constellation Argo Navis, as depicted by Mercator in the mid-16th century. Credit: Public domain.

The space probe *Voyager 2* passed the planets Jupiter, Saturn, Uranus and Neptune. What was the nickname given to its journey across the solar system?

A. The Exploration of the Planets
B. The Grand Tour
C. The New Frontier Tour
D. The Planetary Plan

Which of the following did *not* originally form part of the much larger constellation Argo Navis?

A. Octans
B. Crux
C. Mensa
D. Antila

April 10th

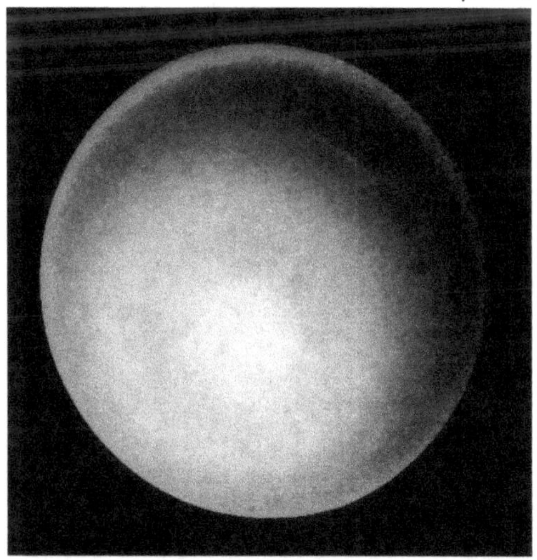

This enhanced Voyager 2 *image of Uranus clearly shows the polar region almost face-on. Credit: NASA/JPL./USGS*

Which planet is often referred to as the "red planet"?

A. Pluto
B. Jupiter
C. Mars
D. Venus

What is the axial tilt of the planet Uranus?

A. 23.5 degrees
B. 44.9 degrees
C. 65.2 degrees
D. 97.8 degrees

April 11th

The Cigar Galaxy, aka The Exploding Galaxy, can be a spectacular sight in photographs. Credit: NASA, ESA, and The Hubble Heritage Team (STScI/AURA).

Which of the following galaxies is *not* in the constellation of Ursa Major?

A. M51 – The Whirlpool Galaxy
B. M81 – Bode's Galaxy
C. M82 – The Cigar Galaxy
D. M101 – The Pinwheel Galaxy

Which planet was the most distant world known to the astronomers of ancient Greece?

A. Jupiter
B. Saturn
C. Uranus
D. Neptune

April 12th

An artist's depiction of sunrise over Janssen, the fourth planet in the 55 Cancri solar system. Credit: Ron Miller/NASAblueshift.

On this date in 1996, astronomers announced the discovery of exoplanet 55 Cancri B. What was so special about it?

A. It was the first Earth-sized planet found
B. It was the most massive exoplanet found
C. It was the first found orbiting a multiple star
D. It had the shortest orbital period of any planet

What was the name of the Greek sun god?

A. Helios
B. Ra
C. Solis
D. Zeus

April 13th

Polaris, the north pole star, never sets for observers in the northern hemisphere. Credit: DSS/Guiseppe Donatiello.

On this date in 2029, which asteroid will pass so close to Earth that it will easily be seen by the naked eye in the night sky?

A. Abbe
B. Amos
C. Apophis
D. Atira

Which constellation is home to Polaris, the star closest to the north celestial pole?

A. Orion
B. Taurus
C. Ursa Major
D. Ursa Minor

April 14th

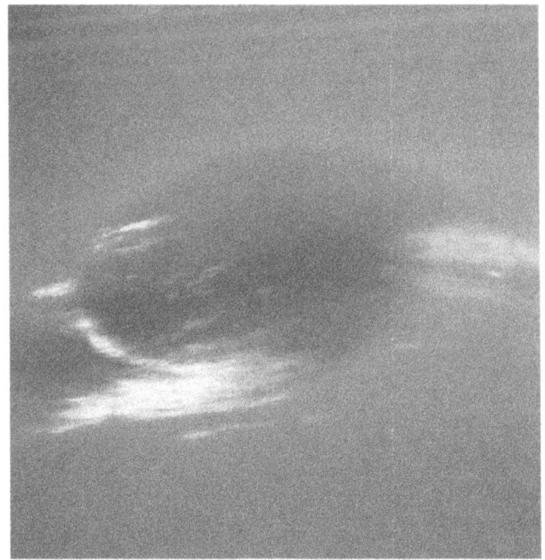

Neptune's Great Dark Spot is similar to the Great Red Spot on Jupiter. Credit: NASA/JPL.

What potential discovery did NASA announce on this date in 2014?

A. An Earth-sized storm in Neptune's atmosphere
B. Flowing water on the surface of Mars
C. A volcano erupting on the surface of Venus
D. A new moon forming in Saturn's A Ring

Which element did Asian astronomers associate with the planet Mercury?

A. Water
B. Air
C. Fire
D. Metal

April 15th

Mizar (upper right) and Alcor (lower left), the famous naked-eye double star of Ursa Major. Credit: Sebastien Lebrigand.

Which two stars in the Big Dipper (or Plough, in the United Kingdom) point toward Polaris, the pole star?

A. Alioth and Dubhe
B. Dubhe and Merak
C. Merak and Phad
D. Mizar and Alcor

Which planet has the greatest variance in surface temperature from day to night?

A. Earth
B. Mars
C. Mercury
D. Pluto

April 16th

This global view of Venus was produced using radar data from NASA's orbiting Magellan *space probe. Credit: NASA.*

Roughly how old is the Sun thought to be?

A. 3.1 billion years old
B. 3.6 billion years old
C. 4.1 billion years old
D. 4.6 billion years old

What's the name given to the southern continent on Venus?

A. Alpha Regio
B. Aphrodite Terra
C. Beta Regio
D. Ishtar Terra

April 17th

The Eta Corvi system is thought to be going through a cometary bombardment, just as our own did. Credit: NASA/JPL-Caltech

About 4 billion years ago the inner solar system went through a period of heavy bombardment from asteroids and comets. What is this period known as?

A. Early Planetesimal Bombardment
B. Late Heavy Bombardment
C. Early Solar Evolutionary Period
D. Mid Planetary Formational Period

What is Pluto's axial tilt?

A. 60 degrees
B. 90 degrees
C. 120 degrees
D. 150 degrees

In this unusual view of the Moon, Mare Orientale is at the center while Mare Procellarum is to the upper right. Credit: NASA/JPL.

Which "sea" (or *mare*) on the Moon is the largest?

A. Mare Procellarum
B. Mare Crisium
C. Mare Imbrium
D. Mare Tranquilitatis

Which constellation covers the greatest area in the night sky?

A. Hydra
B. Orion
C. Ursa Major
D. Virgo

April 19th

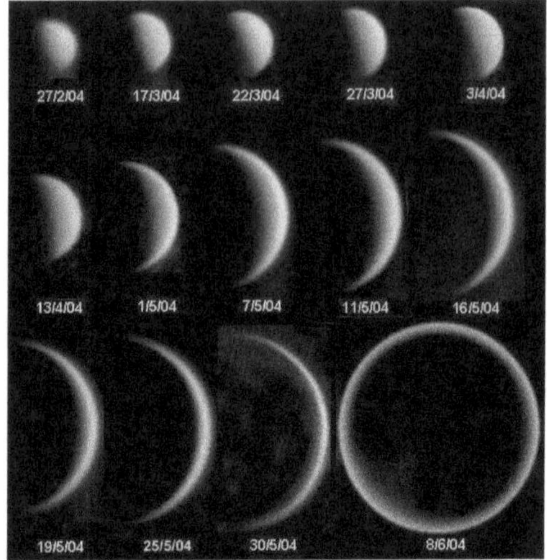

Two of our neighboring worlds show phases like the Moon when viewed through a telescope. Credit: Statis Kalyvas.

Which of the following planets will show phases like the Moon when observed through a telescope?

A. Jupiter
B. Neptune
C. Saturn
D. Venus

How are comets currently named?

A. They are named after a famous scientist
B. They are named after their discoverers
C. The discoverer chooses the name
D. The name is chosen by the International Astronomical Union

April 20th

This image of the Moon's surface was sent back from Surveyor 1, *from its landing site in the Ocean of Storms. Credit: NASA.*

Which space probe was the first to return images from the surface of the Moon in 1967?

A. *Luna 9*
B. *Surveyor 1*
C. *Surveyor 3*
D. *Zond 3*

Which moon of Pluto has a diameter just slightly more than half that of the dwarf planet itself?

A. Charon
B. Hydra
C. Kerberos
D. Nix

April 21st

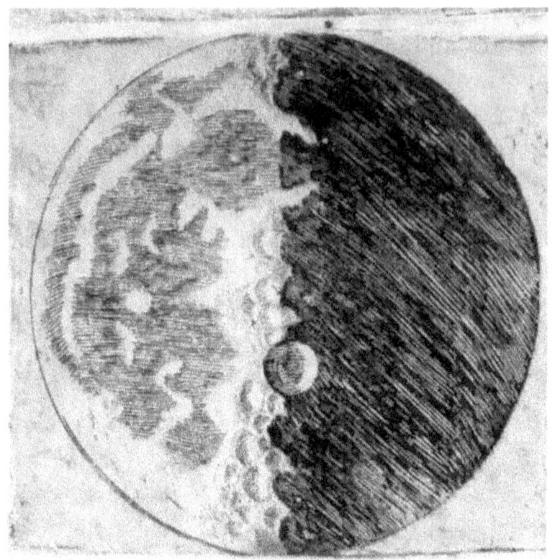

Galileo's sketch of the Moon, as it appeared in his book. Credit: Public Domain.

What was the name of the book Galileo published in 1610 wherein his described his observations with his telescope?

A. *The Harmony of the World*
B. *Mathematical Principles of Natural Philosophy*
C. *On the Revolutions of the Celestial Spheres*
D. *Starry Messenger*

Which famous writer's birth and death both coincided with the appearance of Halley's Comet?

A. Charles Dickens
B. Edgar Allen Poe
C. H. G. Wells
D. Mark Twain

April 22nd

This image of a Leonid meteor was captured at the shower's peak in 2009. Credit: Navicore.

Which meteor shower peaks on this date every year?

A. Orionids
B. Lyrids
C. Virginids
D. Leonids

What's the average temperature on the surface of Mars?

A. -33° C (-27° F)
B. -43° C (-45° F)
C. -53° C (-63° F)
D. -63° C (-82° F)

April 23rd

This image of Jupiter's northern polar region was taken on August 27th, 2016, just two hours before closest approach. Credit: NASA/JPL-Caltech/SwRI/MSSS.

What was the first space probe to take close-up images of Jupiter's north pole?

A. *Galileo*
B. *Juno*
C. *Pioneer 11*
D. *Voyager 2*

What gas makes up roughly 75% of Jupiter's atmosphere?

A. Helium
B. Hydrogen
C. Methane
D. Nitrogen

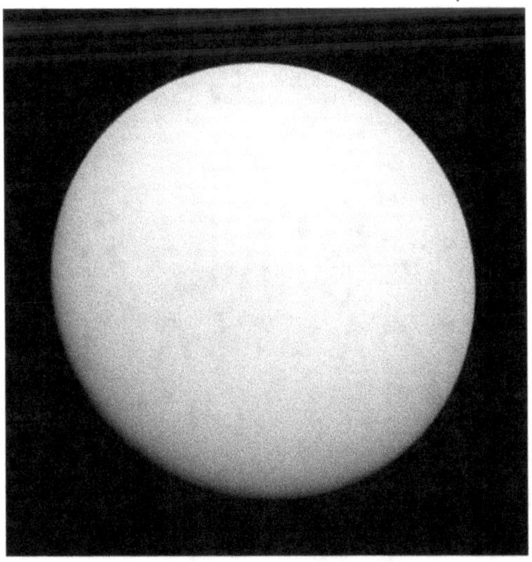

Uranus, as imaged by the Voyager 2 *space probe in January, 1986. Credit: NASA/JPL-Caltech.*

On this date in 2008, astronomers announce the discovery of asteroid 2008 HJ. It has the shortest known rotational period of any asteroid. What is its rotational period?

A. 42 seconds
B. 42 minutes
C. 1 hour, 42 minutes
D. 1 day, 42 minutes

How long is a day on Uranus?

A. 9 hours, 56 minutes
B. 10 hours, 42 minutes
C. 16 hours, 6 minutes
D. 17 hours, 14 minutes

April 25th

Hubble *was released by space shuttle* Discovery. *Credit: NASA/Smithsonian Institution/Lockheed Corporation.*

The *Hubble* Space Telescope was deployed on this date in 1990. What was soon found to be wrong with it?

A. The primary mirror was the wrong shape
B. It was in the wrong orbit
C. The solar panels would not unfold
D. It was not transmitting back data

How many suspected meteor showers are there?

A. 200
B. 400
C. 600
D. 800

April 26th

The craters of the Moon can be a spectacular sight when seen through a small telescope. Credit: Author.

Which planet has the fastest winds?

A. Earth
B. Jupiter
C. Neptune
D. Saturn

Roughly how many craters larger than 1km are thought to be on the visible near side of the Moon?

A. 250,000
B. 300,000
C. 350,000
D. 400,000

April 27th

Mauna Kea, a volcano in the Hawaiian Islands, is home to a number of observatories, including the Keck Telescope. Credit: JPL/Caltech.

What is the most geologically active world in the solar system?

A. Earth
B. Io (Jupiter)
C. Triton (Neptune)
D. Venus

What is the approximate tilt of the Earth's axis?

A. 23 degrees
B. 25 degrees
C. 27 degrees
D. 29 degrees

April 28th

This artist's impression depicts Chariklo, the largest known Centaur asteroid and the first to be discovered with rings. Credit: ESO/L. Calçada/Nick Risinger.

How long did it take Clyde Tombaugh to discover Pluto?

A. Four weeks
B. About six months
C. Just under a year
D. A little more than two years

Where do the Centaur asteroids orbit the Sun?

A. Within the orbit of Mercury
B. Between Mars and Jupiter
C. Between the outer planets Jupiter and Neptune
D. Beyond the orbit of Neptune

April 29th

This time-lapse photo shows the stars as they spin about Polaris, the north pole star. Credit: Q-lieb-in (from Wikimedia).

What chemical gives some meteors a yellow color?

A. Iron
B. Magnesium
C. Potassium
D. Sodium

What does it mean if a star or constellation is said to be circumpolar?

A. The stars never rise from the observer's location
B. The stars never set from the observer's location
C. The stars pass directly overhead
D. The stars are only visible from the northern hemisphere

April 30th

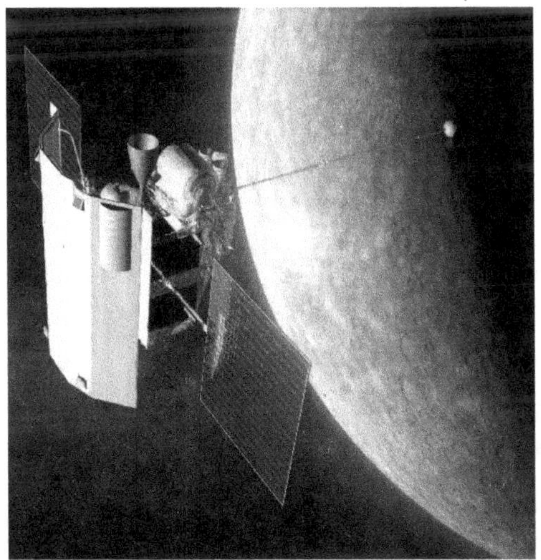

MESSENGER *orbited Mercury for four years, from 2011 to 2015. Credit: NASA/JHU/APL.*

What happened to the *MESSENGER* space probe at the end of its mission to Mercury on this date in 2015?

A. It crashed into the surface of Mercury
B. Its batteries died but it still orbits the planet
C. It went on to orbit Venus
D. It returned to Earth

Which of the following planets has a day that lasts longer than its year?

A. Jupiter
B. Mars
C. Venus
D. Neptune

May

May 1st

Tethys is Saturn's fifth largest moon and was discovered by Giovanni Cassini in 1684. Credit: NASA/JPL-Caltech/Space Science Institute.

In what year was Nereid, Neptune's second moon, discovered?

A. 1934
B. 1939
C. 1944
D. 1949

What's the name of the 400km crater that dominates the surface of Saturn's moon Tethys?

A. Hipparchos
B. Homer
C. Odysseus
D. Plato

The Roman god Mercury, as imagined by the artist Hendrik Goltzius in 1611. Credit: Public Domain.

In which city did the International Astronomical Union hold its inaugural meeting on this date in 1922?

A. London
B. Paris
C. Rome
D. Washington D.C.

Mercury was named after a Roman god but who was his Greek equivalent?

A. Hermes
B. Ares
C. Demeter
D. Apollo

May 3rd

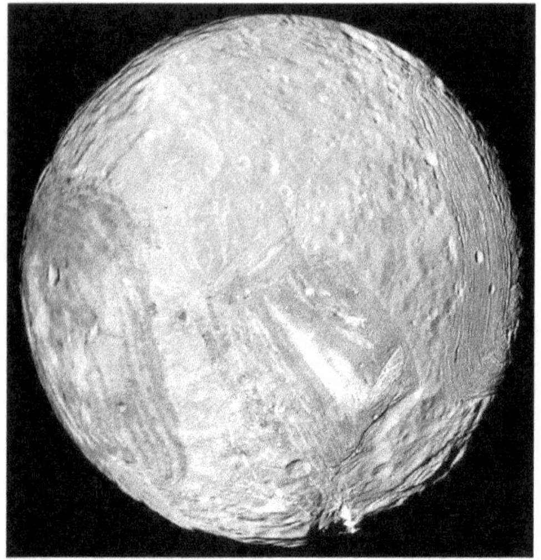

This image of Miranda was taken by the Voyager 2 *space probe in 1986. Credit: NASA/JPL-Caltech.*

The moon Miranda has a feature known as the Verona Rupes, a 5-10 km high scarp that are the tallest cliffs in the solar system. Which planet does the moon orbit?

A. Jupiter
B. Neptune
C. Saturn
D. Uranus

What's the only moon in the solar system known to possess a magnetosphere?

A. The Moon (Earth)
B. Ganymede (Jupiter)
C. Titan (Saturn)
D. Triton (Neptune)

May 4th

This image shows the size of our Sun in comparison to the other planets in the solar system. Credit: Lsmpascal (from Wikimedia).

Which astronomer was the first to report active volcanoes on the Moon in 1783?

A. William Herschel
B. Giovanni Maraldi
C. Edmond Halley
D. Charles Messier

How much more massive is the Sun compared to the Earth?

A. 330,000x
B. 360,000x
C. 390,000x
D. 420,000x

May 5th

An artist's rendering of the Cassini *space probe observing a sunset through Titan's hazy atmosphere. Credit: NASA/JPL-Caltech.*

Which planet has clouds of sulfuric acid in the sky?

A. Jupiter
B. Mars
C. Uranus
D. Venus

What gas makes up 98% of Titan's atmosphere?

A. Hydrogen
B. Methane
C. Nitrogen
D. Oxygen

In this view of the Earth, taken January 4th 2012, the North American continent is clearly visible. Credit: NASA/NOAA/GSFC/Suomi NPP/VIIRS/Norman Kuring

What is the precise length of a year on Earth?

A. 364.89 days
B. 365.26 days
C. 366.56 days
D. 367.29 days

What is Saturn's moon Tethys primarily composed of?

A. Iron
B. Methane ice
C. Rock
D. Water ice

May 7th

This painting shows Antares (right), a red supergiant, with its companion star. Credit: Sephirohq (from Wikimedia.).

Which star's name means "rival of Mars"?

A. Aldebaran
B. Antares
C. Arcturus
D. Betelgeuse

Which gas makes up nearly 97% of Venus' atmosphere?

A. Carbon dioxide
B. Methane
C. Nitrogen
D. Sulphur dioxide

Saturn's moon Rhea appears against Titan. Credit: NASA/JPL-Caltech/Space Science Institute.

Which moon is the only one covered in an atmosphere that hides the surface from view?

A. Enceladus (Saturn)
B. Europa (Jupiter)
C. Titan (Saturn)
D. Triton (Neptune)

Many believe the Moon formed from the debris of a collision between the proto-Earth and a Mars-sized body. What's the name given to that theoretical body?

A. Anubis
B. Markab
C. Nibiru
D. Theia

May 9th

Sunlight does not arrive instantly but takes time to reach the Earth. Credit: NASA.

How long does it take light to travel from the Sun to the Earth?

A. 8 seconds
B. 8 minutes and 20 seconds
C. 8 hours and 20 minutes
D. 8 days and 20 hours

Why can't the *Hubble* Space Telescope study Mercury?

A. Hubble isn't powerful enough
B. Mercury isn't visible from Earth orbit
C. The optics will be damaged by the nearby Sun
D. The planet is too small

May 10th

In this radar image of Venus, Ishtar Terra can be seen as the north polar region of the planet. Credit: NASA/JPL/USGS

If you're looking at the zenith, what part of the sky are you looking at?

A. Directly overhead
B. Due north
C. Due south
D. Midway between the horizon and overhead

The northern continent on Venus is called Ishtar Terra. Who was Ishtar?

A. The Babylonian goddess of love
B. The Egyptian goddess of the sky
C. The Greek goddess of war
D. The Roman Earth mother

May 11th

The Sombrero Galaxy, in the constellation of Virgo. Credit: ESO/IDA/Danish 1.5 m/R. Gendler and J.-E. Ovaldsen.

Who discovered Messier 104, the Sombrero Galaxy, on this date in 1781?

A. William Herschel
B. Jean-Dominique Maraldi
C. Pierre Méchain
D. Charles Messier

Which astronomer was initially co-credited with the discovery of Neptune but had the honor removed late in the 20th century?

A. Galileo Galilei
B. John Flamsteed
C. John Couch Adams
D. Percival Lowell

May 12th

This artist's impression depicts Pluto's largest moon rising over the dwarf planet's frozen south pole. Credit: NASA/JHUAPL/SwRI.

Which planet is the furthest from the Sun?

A. Jupiter
B. Mercury
C. Neptune
D. Pluto

If you could stand on the surface of Pluto, how large would its largest moon Charon appear in the sky?

A. About the size of the full Moon
B. About 3x larger than the full Moon
C. About 5x larger than the full Moon
D. About 8x larger than the full Moon

May 13th

Comet Donati appeared close to the bright star Arcturus on October 5th, 1858. Credit: E. Weiß.

Which star's name means "guardian of the bear"?

A. Arcturus
B. Mizar
C. Polaris
D. Vega

What is Neptune's axial tilt?

A. Zero degrees
B. 14 degrees
C. 28 degrees
D. 42 degrees

May 14th

The Griffith Observatory is a famous landmark in the Los Angeles area. Credit: Author.

In which year was the Griffith Observatory in Los Angeles opened to the public on this date?

A. 1935
B. 1940
C. 1945
D. 1950

Discrepancies in the predicted position of the planet Uranus led to the discovery of which other world?

A. Haumea
B. Neptune
C. Pluto
D. Sedna

May 15th

The New Horizons *probe obtained this image of Jupiter's rings while en-route to Pluto. Credit: NASA/Johns Hopkins University Applied Physics Laboratory/Southwest Research Institute.*

In which year did the International Astronomical Union officially define the 88 known constellations?

A. 1897
B. 1922
C. 1947
D. 1972

What are Jupiter's rings comprised of?

A. Dust
B. Ice
C. Rock
D. Ice/Rock combination

May 16th

Mons Huygens (upper right of center) is the tallest mountain on the Moon. Credit: James Stuby from an original NASA image.

In Greek mythology, which constellation is *not* associated with one of the 13 labors of Hercules?

A. Cancer, the Crab
B. Draco, the Dragon
C. Leo, the Lion
D. Ursa Major, the Great Bear

What is the base to peak height of Mons Huygens, the tallest mountain on the Moon?

A. 2.7 km (1.7 miles)
B. 4.0 km (2.5 miles)
C. 5.5 km (3.4 miles)
D. 6.9 km (4.3 miles)

May 17th

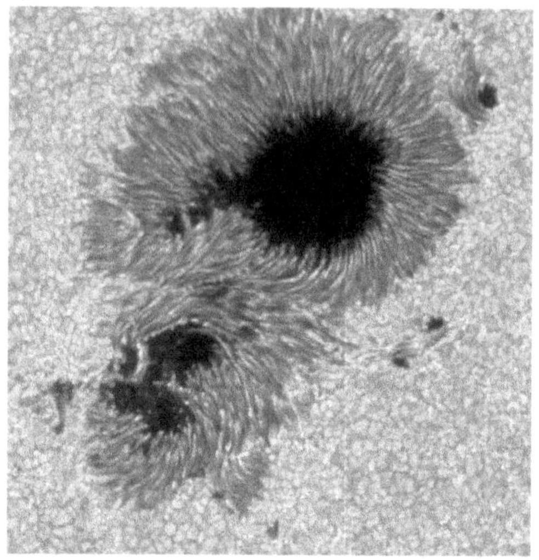

Sunspots appear black against the Sun's surface. The largest one in this image is bigger than the Earth. Credit: NASA/JAXA.

Why do sunspots appear black against the Sun's surface?

A. Their chemical composition is different
B. They are clouds in the sun's atmosphere
C. They are cooler than the surrounding surface
D. Nobody knows for sure

What rare astronomical event was observed for the first time on this date in 1737?

A. Venus occulting (covering) Mercury
B. A meteor storm of thousands of shooting stars
C. The Moon occulting Uranus
D. The zodiacal light

May 18th

This image of Halley's Comet was taken on May 29th, 1910, from Yerkes Observatory. Credit: Public Domain

Which bird did Mayan astronomers associate with the planet Mercury?

A. Eagle
B. Hawk
C. Kestrel
D. Owl

Which amateur astronomer created the Caldwell Catalog of deep sky objects?

A. Caroline Herschel
B. Patrick Moore
C. Cathy Caldwell
D. Carl Sagan

May 19th

The crescent Earth rising over the limb of the Moon, as imaged by the crew of Apollo 17. *Credit: NASA.*

The Earth passed through the tail of Halley's Comet on this date in 1910. Why did doomsayers predict the end of life on Earth as a result?

A. The tail contained elements of poisonous gas
B. The comet was predicted to hit the Earth
C. The comet was predicted to hit the Sun
D. Comet dust would rain down upon the Earth

Who is credited with first theorizing that the Earth was a sphere?

A. Aristotle
B. Copernicus
C. Galileo
D. Pythagoras

May 20th

The constellation Crux (the Southern Cross) is a prominent and familiar sight to observers in the southern hemisphere. Credit: Naskies (from Wikimedia)

Which constellation covers the smallest area in the night sky?

A. Canis Minor
B. Crux
C. Delphinus
D. Vulpecula

How many main rings does Neptune have?

A. Three
B. Five
C. Seven
D. Nine

May 21st

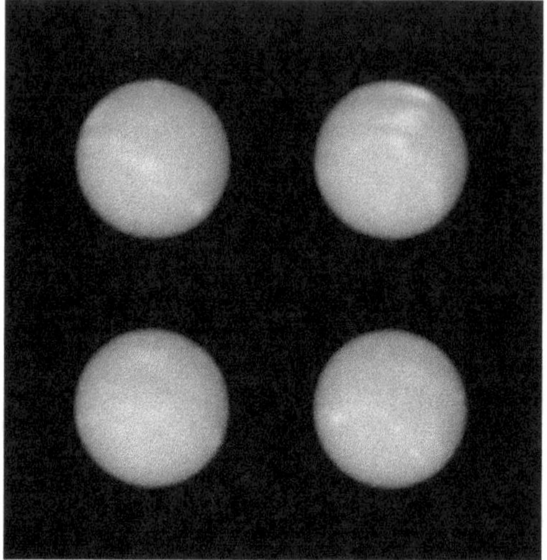

These four images from the Hubble *Space Telesope show Neptune rotating over a 16 hour period. Credit: NASA, ESA, and the Hubble Heritage Team (STScI/AURA)*

What's the name given to the path the Sun, Moon and planets appear to take as they move through the zodiac?

A. Arc
B. Trajectory
C. Orbit
D. Ecliptic

What's the minimum equipment needed to find Neptune?

A. No equipment needed – visible to the naked eye
B. Binoculars
C. Small Telescope
D. Large Telescope

May 22nd

The Earth/Moon system is sometimes called a "double planet" because of the size of the Moon in comparison with the Earth. Credit: NASA.

What's the only planet bright enough to be visible in the daytime sky?

A. Jupiter
B. Mars
C. Mercury
D. Venus

How large is the Moon compared to the Earth?

A. A third the size of Earth
B. A quarter the size of Earth
C. A tenth the size of Earth
D. A twentieth the size of Earth

May 23rd

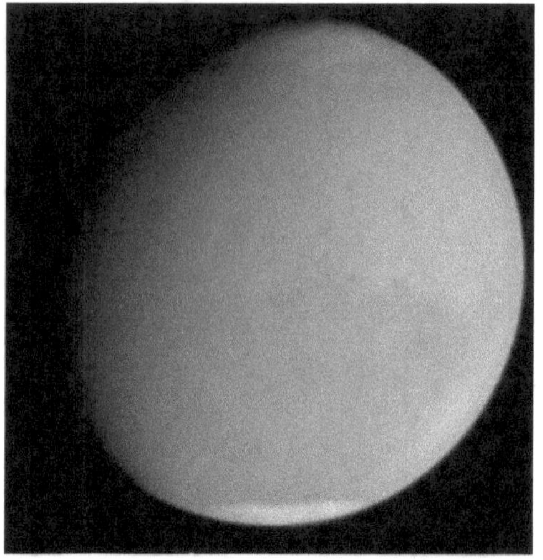

This world experiences global dust storms – but which world is it? Credit: NASA

Venus was named after a Roman god but who was his Greek equivalent?

A. Aphrodite
B. Ceres
C. Hathor
D. Rhea

Which planet or moon has the largest dust storms in the solar system that can cover the entire planet?

A. Titan (Saturn)
B. Venus
C. Earth
D. Mars

Proteus is one of the tiny moons of Neptune discovered by the Voyager 2 space probe in 1989. Credit: NASA

How many moons of Neptune were known to exist and named before *Voyager 2* passed the planet in August, 1989?

A. Two
B. Four
C. Six
D. Eight

The Chinese zodiac had its origins with the observed movement of which planet across the sky?

A. Jupiter
B. Mars
C. Venus
D. Saturn

May 25th

This painting depicts a total eclipse of the Sun by the Earth, as seen from the Moon. Credit: Lucien Rudaux

There are twelve officially recognized constellations in the zodiac but the Sun, Moon and planets also appear to move through which other constellation?

A. Andromeda
B. Canis Major
C. Ophiuchus
D. Ursa Major

In what year, on this date, did the first predicted solar eclipse occur?

A. 585 B.C.
B. 23 A.D.
C. 1167 A.D.
D. 1623 A.D.

May 26th

Uranus's moon Titania was discovered by William Herschel on January 11th, 1787. Credit: NASA/JPL.

What is Uranus' largest moon?

A. Ariel
B. Oberon
C. Titania
D. Umbriel

Which planet would theoretically float in water if you could find an ocean large enough?

A. Jupiter
B. Saturn
C. Uranus
D. Neptune

May 27th

This artist's impression shows asteroids surrounding the nearby star Epsilon Eridani. Credit: NASA/JPL-Caltech.

Where do the majority of asteroids orbit the Sun?

A. Between Earth and Mars
B. Between Mars and Jupiter
C. Between Jupiter and Saturn
D. Beyond the orbit of Saturn

Light from which star was used to open Chicago World's Fair on this date in 1933?

A. Arcturus
B. Sirius
C. Spica
D. Vega

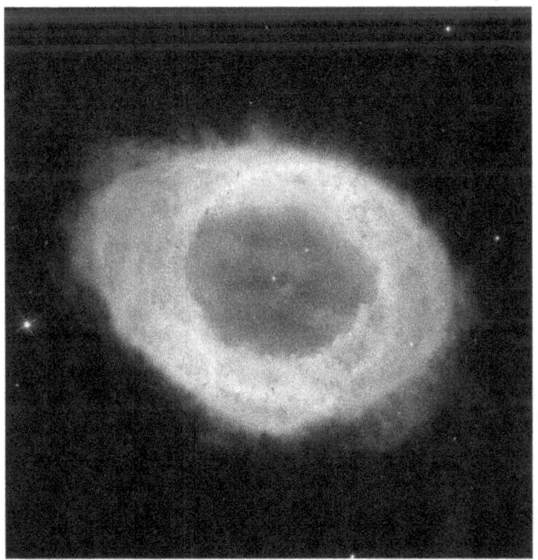

The Ring Nebula lies over 2,000 light years away and looks like a tiny smoke ring through a telescope. Credit: NASA, ESA, and C. Robert O'Dell (Vanderbilt University).

In what year did J. F. Encke discover the gap in Saturn's rings that would later bear his name?

A. 1822
B. 1827
C. 1832
D. 1837

Which of the following cannot be found in the constellation of Sagittarius?

A. The Lagoon Nebula
B. The Ring Nebula
C. The Swan Nebula
D. The Trifid Nebula

May 29th

Cor Caroli is the brightest star in the constellation of Canes Venatici, the Hunting Dogs. Credit: Sidney Hall, 1825.

On this date in 1660, what event was said to cause the star Cor Caroli ("Charles's Heart") to shine especially brightly?

A. The birth of King Charles II
B. The coronation of King Charles II
C. The return of King Charles II to England
D. The death of King Charles II

Which planet has an atmospheric surface pressure 92 times greater than the Earth's?

A. Mars
B. Mercury
C. Pluto
D. Venus

May 30th

This image demonstrates Jupiter's immense size in comparison to the Earth. Credit: NASA.

If the planet Jupiter were hollow, how many Earths could you fit inside it?

A. 1.045
B. 1,321
C. 1,620
D. 1,925

Saturn's moon Titan accounts for how much of the mass in orbit about the planet?

A. 69%
B. 76%
C. 83%
D. 96%

May 31st

Io is Jupiter's third largest moon and the most volcanic body in the solar system. Credit: NASA/JPL/University of Arizona

On a star chart, what's the astronomical equivalent of latitude called?

A. Altitude
B. Declination
C. Latitude
D. Right Ascension

What is the primary component of the thin atmosphere that surrounds Jupiter's moon Io?

A. Carbon Dioxide
B. Nitrogen Oxide
C. Oxygen
D. Sulphur dioxide

June

June 1st

Mare Crisium can easily be seen as the oval patch close to the upper right edge of the Moon. Credit: NASA

What's the largest crater on the Moon?

A. Aitken Basin at the South Pole
B. Copernicus
C. Mare Crisium
D. Tycho

What is thought to be at the Earth's core?

A. Iron
B. Mercury
C. Sulphur
D. The land before time

June 2nd

Ganymede is Jupiter's largest moon and one of the four Galilean satellites. Credit: NASA/JPL.

Which of the five Chinese elements is associated with Saturn?

A. Air
B. Earth
C. Fire
D. Water

Who suggested the names given to the Galilean satellites of Jupiter?

A. Tycho Brahe
B. Nicolaus Copernicus
C. Galileo Galilei
D. Johannes Kepler

June 3rd

The waning crescent Moon photographed by the Expedition 24 *crew of the* International Space Station. *Credit: NASA.*

Neptune was named after a Roman god but who was his Greek equivalent?

A. Aquarius
B. Nereus
C. Oceanus
D. Poseidon

What does it mean if the Moon is said to be waning?

A. The Moon is growing fuller
B. The Moon appears to be growing thinner
C. The Moon is rising in the morning
D. The Moon is settng in the evening

Most of the nearby stars are red dwarves, as depicted in this artist's impression. Credit: NASA/Harvard-Smithsonian Center for Astrophysics/D. Aguilar.

Which star is the closest to the Sun?

A. Polaris
B. Proxima Centauri
C. Sirius
D. Vega

How much more massive is Uranus compared to the Earth?

A. 5x
B. 10x
C. 15x
D. 20x

June 5th

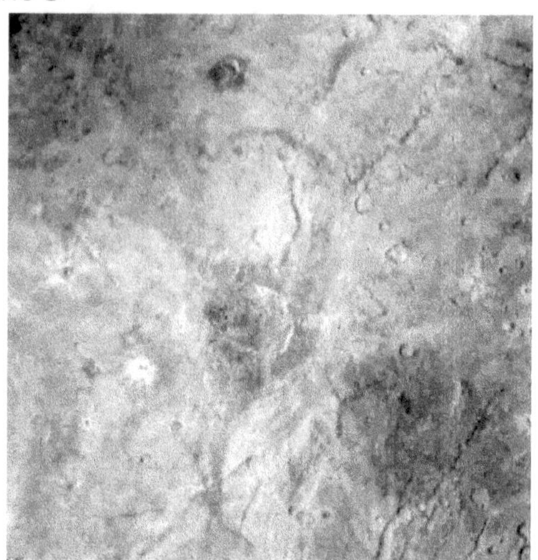

This close-up view of Charon's surface was taken by the New Horizons *space probe. Credit: NASA/JPL.*

What is Pluto's largest moon?

A. Charon
B. Hydra
C. Nix
D. Triton

What's the average distance from Jupiter to the Sun? (1 AU = the average distance from the Earth to the Sun.)

A. 3.9 AU
B. 5.2 AU
C. 7.5 AU
D. 9.1 AU

Future astronauts on Mars encounter the long-dead Viking 2 *lander probe. Credit: NASA/Pat Rawlings*

What is Saturn's largest moon?

A. Enceladus
B. Mimas
C. Rhea
D. Titan

How strong is the gravity on Mars compared to the Earth?

A. 23%
B. 28%
C. 33%
D. 38%

June 7th

This image of Pluto was captured by the New Horizons *space probe on July 14th, 2015. Credit: NASA/JHUAPL/SwRI.*

Where in Arizona did Percival Lowell build the observatory that now bears his name?

A. Flagstaff
B. Mesa
C. Phoenix
D. Tucson

What's the official name for the heart-shaped feature easily seen in photographs of Pluto?

A. Corazón Regio
B. Lowell Regio
C. Tombaugh Regio
D. Valentino Regio

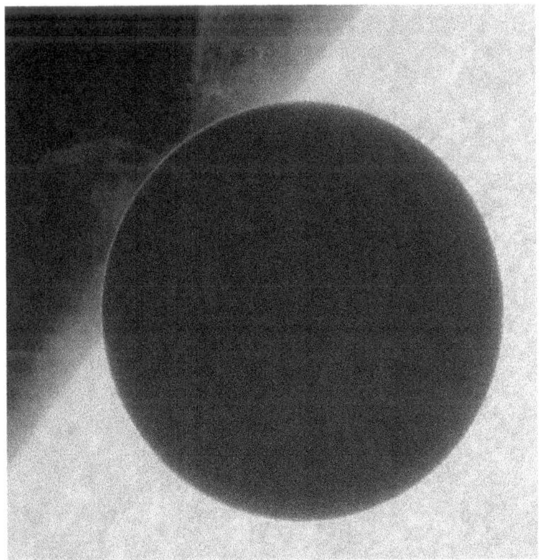

This stunning image of Venus transiting the Sun was captured by Japan's Hinode *space observatory on June 6th, 2012. Credit: JAXA/NASA.*

In what year did the first of four 21st century transits of Venus across the Sun occur?

A. 2000
B. 2004
C. 2008
D. 2012

What does the asteroid Pallas have in common with the planet Uranus?

A. It orbits on its side
B. It has a dense core
C. It has a system of faint rings
D. It's the same distance from the Sun

June 9th

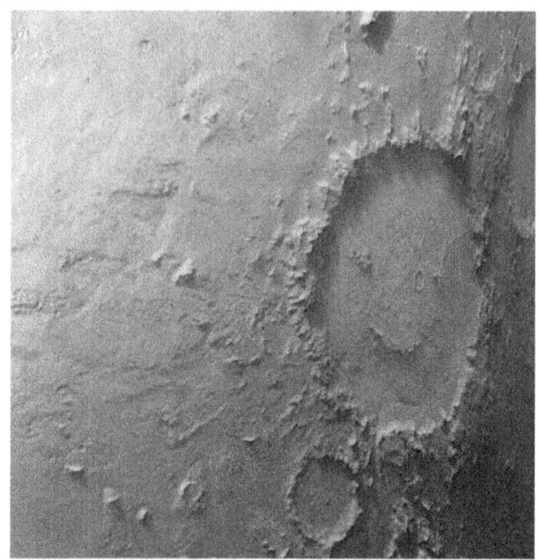

Galle Crater on Mars, also known as the "Happy Face" crater. Credit: NASA/JPL/MSSS

What's the name given to the area of sky from which meteors appear?

A. The foundation
B. The point of origin
C. The radiant
D. The source

What, or whom, are craters on Mars smaller than 60km named for?

A. Capital cities of the world
B. Deceased world leaders
C. Science fiction writers
D. Towns with less than 100,000 people

June 10th

This artist's impression depicts the Milky Way as it might appear from above. Credit: Nick Risinger

Which astronomer was the first to predict the return of a comet?

A. Galileo Galileo
B. Edmond Halley
C. Johannes Kepler
D. William Herschel

How many stars are thought to be in our own Milky Way galaxy?

A. Between 25 and 50 billion
B. Between 50 and 75 billion
C. Between 75 and 100 billion
D. More than 100 billion

June 11th

An artist's impression of the New Horizons *probe (lower right) encountering a Kuiper Belt object beyond the orbit of Neptune. Credit: JHUAPL/SwRI.*

On this date in 2008 the International Astronomical Union announced a new name for trans-Neptunian dwarf planets. What is it?

A. Asteroids
B. Helioids
C. Kuiperoids
D. Plutoids

How long is a day on Jupiter?

A. 9 hours, 56 minutes
B. 10 hours, 39 minutes
C. 11 hours, 18 minutes
D. 12 hours, 47 minutes

The Great Lakes of North America can be clearly seen in this image taken by the space shuttle crew of STS-84 *in 1997. Credit: NASA.*

When ideally placed in the sky, when is the best time of year to observe Mars?

A. Late January to late March
B. Late April to late June
C. Late July to late September
D. Late October to late December

What percentage of the Earth's water is fresh rather than salt water?

A. 2.5%
B. 5%
C. 7.5%
D. 10%

June 13th

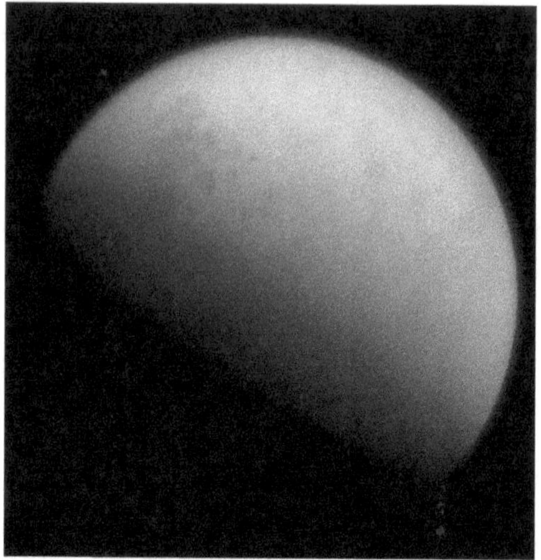

Saturn's moon Titan, as imaged by the Cassini *space probe in 2013. Credit: NASA/JPL-Caltech/Space Science Institute.*

How did Christian Huygens announce his discovery of Saturn's largest moon Titan in 1655?

A. As an anagram
B. At an astronomical meeting
C. In a letter to a peer
D. In a scientific publication

In which month is the Earth furthest from the Sun?

A. January
B. April
C. July
D. October

This close-up and detailed view of Messier 63, the Sunflower Galaxy, was taken in 2015 by the Hubble Space Telescope. *Credit: NASA.*

Which galaxy did the French astronomer Pierre Méchain discover in 1779?

A. Andromeda Galaxy
B. Blackeye Galaxy
C. Sunflower Galaxy
D. Whirlpool Galaxy

Which moon of Neptune did not have an official name until the 20th century?

A. Larissa
B. Nereid
C. Proteus
D. Triton

June 15th

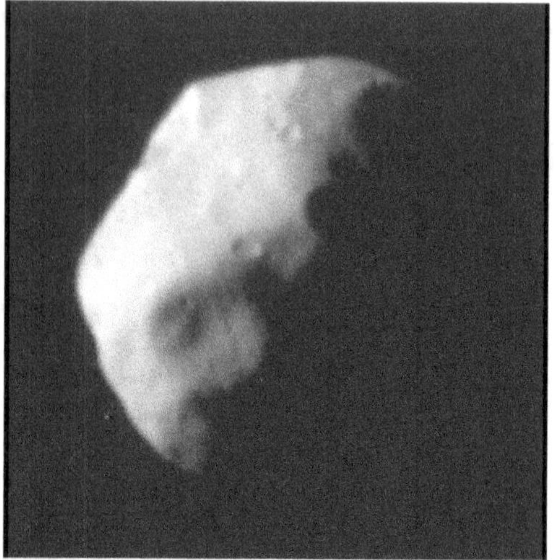

Pluto's moon Nix, as imaged by the New Horizons *space probe. Credit: NASA/JHUAPL/SwRI.*

Hydra and Nix, two new moons of Pluto, are discovered in images from the *Hubble* Space Telescope on this date in 2005. Including the two new moons, how many moons was Pluto known to have at that time?

A. Two
B. Three
C. Four
D. Five

What percentage of sunlight is made up of visible light?

A. 20%
B. 40%
C. 60%
D. 80%

June 16th

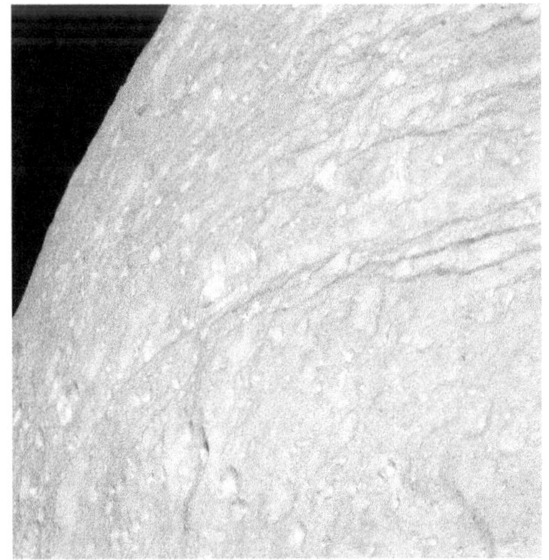

Cassini took this detailed image of Tethys in 2005. The chasm that almost encircles the moon can be seen across the center. Credit: NASA/JPL/Space Science Institute.

How many moons does Mercury have?

A. None
B. One
C. Two
D. Three

What's the name of the canyon that extends 270° around the surface of Saturn's moon Tethys?

A. Catalogus Plantarum
B. Ithaca Chasma
C. Marineris Chasma
D. Utica Valleris

June 17th

A total lunar eclipse can be an awe-inspiring sight. Credit: Radoslaw Ziomber.

On this date in 1985 the *Vega 2* space probe flew by Venus on its way to an encounter with Comet Halley. What did it release into the planet's atmosphere?

A. A balloon
B. An atmospheric probe
C. Microbes
D. A lander

What's the name given to the cycle of lunar and solar eclipses?

A. Espenak
B. Luminaris
C. Saros
D. Solis

June 18th

The open star cluster Messier 6 can be found in Scorpius, the Scorpion. Credit: Author

On this date in 1178 five monks witness an apparent impact upon the Moon. What crater is thought to have formed as a result?

A. Abbot
B. Copernicus
C. Giordano Bruno
D. Julius Caesar

Which character from Greek mythology was killed by Scorpius, the Scorpion?

A. Cepheus, the King
B. Hercules, the Hero
C. Orion, the Hunter
D. Perseus the Hero

June 19th

This logo was produced in 1974 to commemorate the Viking *missions to Mars. Credit: NASA/ Langley Research Center*

Who made the first known published drawing of Saturn in 1633?

A. Giovanni Cassini
B. Pierre Gassendi
C. Galileo Galilei
D. Christiaan Huygens

Viking 1 entered orbit about Mars on this date but in which year?

A. 1973
B. 1974
C. 1975
D. 1976

June 20th

This artist's depiction imagines what Venus might look like if it were terraformed to support life. Credit: Ittiz (from Wikimedia).

The dark patches easily seen on the face of the Moon are known are *mare*. What is the word *mare* latin for?

A. Dream
B. Mirage
C. Ocean
D. Plain

Which planet is sometimes described as being Earth's twin?

A. Jupiter
B. Mercury
C. Saturn
D. Venus

June 21st

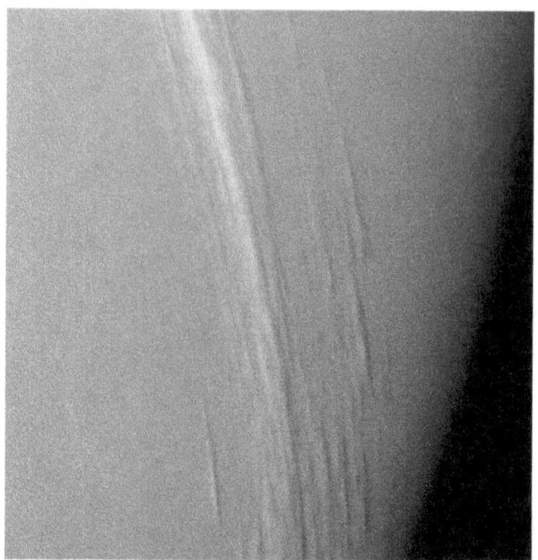

This image shows clouds high in the atmosphere of Neptune. It was taken by the Voyager 2 *space probe just two hours before closest approach. Credit: NASA/JPL*

The summer solstice occurs around this date in the northern hemisphere. What is the event's official name?

A. Autumnal equinox
B. Estival solstice
C. Hibernal solstice
D. Vernal equinox

What's the fastest wind speed recorded in Neptune's atmosphere?

A. 1,800 kph (1,100 mph)
B. 2,100 kph (1,300 mph)
C. 2,400 kph (1,500 mph)
D. 2,700 kph (1,700 mph)

It might look like Saturn in this image, but this ultraviolet view is of the planet Uranus, its rings and its five largest moons. Credit: ESO.

What did James W. Christy discover on this date in 1978?

A. The rings of Uranus
B. A supernova
C. Pluto's largest moon, Charon
D. Active volcanos on Jupiter's moon Io

Who became the first Astronomer Royal on this date in 1675?

A. John Flamsteed
B. Edmond Halley
C. William Herschel
D. Isaac Newton

June 23rd

This artist's impression shows a crescent Makemake lit by distant sunlight. Credit: IAU, M. Kornmesser (ESA/Hubble).

Which dwarf planet is the largest?

A. Ceres
B. Eris
C. Makemake
D. Pluto

The constellation of Boötes, the Herdsman, resembles what shape?

A. A diamond
B. A kite
C. A rectangle
D. A stick figure

Cassini *captured this view of Jupiter and its moon Io as it passed the planet. Credit: NASA/JPL/University of Arizona*

Which of the following is not considered to be a summer constellation?

A. Aquila
B. Cepheus
C. Cygnus
D. Lyra

Which planet did the space probe *Cassini* pass on this date in 1999 while en-route to Saturn?

A. Earth
B. Jupiter
C. Mercury
D. Venus

June 25th

The Big Dipper (lower left) and Polaris (upper right), the Pole Star, are prominently featured on the flag of Alaska. Credit: Public Domain.

Which of the following is an inferior planet?

A. Mercury
B. Mars
C. Jupiter
D. Saturn

Which constellation contains the seven bright stars known as the Big Dipper in North America or the Plough in the United Kingdom?

A. Andromeda
B. Orion
C. Ursa Major
D. Ursa Minor

June 26th

The Keystone Cluster contains about 300,000 stars and lies approximately 22,000 light years away. Credit: Author.

Which star represents the red heart of Scorpius, the Scorpion?

A. Aldebaran
B. Antares
C. Arcturus
D. Betelgeuse

In which constellation can the Keystone Cluster be found?

A. Hercules
B. Ophiuchus
C. Orion
D. Ursa Major

June 27th

Saturn's rings are not solid but are actually clumps of water ice. Credit: NASA/JPL/University of Colorado.

How thick are the rings of Saturn?

A. About 9 meters (30 feet)
B. About 49 meters (160 feet)
C. About 88 meters (290 feet)
D. About 128 meters (420 feet)

The asteroid Icarus was discovered on this date in 1949. What was unique about the asteroid at that time?

A. It had the shortest rotational period of any known asteroid
B. It was the smallest asteroid at that time
C. It passed closer to the Sun than any other asteroid
D. It was the densest known asteroid

Saturn's moons Epimetheus (left) and Janus (right) are co-orbital. Credit: NASA/JPL/Space Science Institute.

Which of the following was the first NASA space probe to fly past more than one planet?

A. *Galileo*
B. *Mariner 10*
C. *Pioneer 10*
D. *Voyager 1*

What are co-orbital satellites?

A. Two moons are the remnants of a single moon
B. Two moons whose orbits cross
C. Two moons on opposite sides of the same planet
D. Two moons that share the same orbit

June 29th

This image shows the center of the Milky Way rising into the skies above Paranal, Chile. Credit: ESO/B. Tafreshi (twanight.org).

What does the word "asteroid" mean?

A. Little moon
B. Little planet
C. Little star
D. Little sun

In which constellation can you find the center of our own Milky Way galaxy?

A. Cassiopeia
B. Cygnus
C. Orion
D. Sagittarius

This image from 1929 shows the devastation caused by a suspected asteroid explosion near the Stony Tunguska River in Russia. Credit: Leonid Kulik.

The United Nations has declared this date to be an annual event to educate the public on which of the following?

A. Asteroids
B. Comets
C. Eclipses
D. Planet X

In what year on this date did a suspected asteroid explosion flatten trees near the Stony Tunguska River?

A. 1906
B. 1908
C. 1910
D. 1912

July

July 1st

This artist's impression shows the Cassini *space probe as it orbits Saturn. Credit: NASA/JPL.*

In what year did the *Cassini* space probe become the first spacecraft to orbit Saturn?

A. 2002
B. 2003
C. 2004
D. 2005

What's the core of a comet known as?

A. The coma
B. The epicenter
C. The nucleus
D. The plasma center

July 2nd

The waxing crescent Moon and Venus, taken just minutes before an occultation. Credit: Vinish K Saini.

How long does it take Venus to complete one orbit of the Sun?

A. 204.3 days
B. 224.7 days
C. 244.2 days
D. 264.9 days

What is an occultation?

A. When one object completely covers another
B. When an object appears directly overhead
C. When the full Moon occurs at the same time as a solar eclipse
D. When a planet appears opposite the Sun in the sky

July 3rd

The heart of NGC 6101, a globular cluster in the southern constellation of Apus. Credit: NASA/HST.

How long does it take Mercury to complete one orbit of the Sun?

A. 46 days
B. 88 days
C. 122 days
D. 146 days

What does the modern constellation Apus represent?

A. A bird of paradise
B. An eagle
C. An owl
D. A parrot

A rock painting by the Anasazi Native Americans depicting the supernova of 1054 A.D. Credit: Alex Marentes.

In what year did Chinese astronomers observe the supernova that would later become the Crab Nebula?

A. 1039
B. 1044
C. 1049
D. 1054

On this date in 1997 the *Pathfinder* probe landed on the surface of Mars. What was the name of the robotic rover it carried?

A. Curiosity
B. Opportunity
C. Sojourner
D. Spirit

July 5th

Juno *took this image of Jupiter's atmosphere from a distance of only 14,500 km (9,000 miles). Credit: NASA/JPL-Caltech/SwRI/MSSS/Roman Tkachenko.*

Phobos and Deimos are the two small moons of Mars. In mythology, what relationship did they have to Ares, the Greek equivalent of Mars?

A. Brothers
B. Father and Grandfather
C. Sons
D. Uncles

In what year did the space probe *Juno* arrive at Jupiter?

A. 2014
B. 2015
C. 2016
D. 2017

Planets are thought to form from collisions between comets, asteroids and other proto planets. Credit: NASA/JPL-Caltech

Saturn was named after a Roman god but who was his Greek equivalent?

A. Cronus
B. Hermes
C. Set
D. Titan

Roughly how old is the Earth?

A. 3.2 billion years old
B. 4.5 billion years old
C. 5.7 billion years old
D. 6.1 billion years old

July 7th

This artist's impression shows the ring of debris thought to surround the star Vega. Credit: NASA/JPL-Caltech.

On this date in 1988, the Soviet Union launched the *Phobos 1* probe to Mars but the probe failed due to a computer malfunction en-route. What was the cause of the malfunction?

A. The ground control computer had a virus
B. Ground control systems were hacked
C. Bad computer hardware
D. Bad software uploaded by ground control

Which star is *not* a part of the Summer Triangle?

A. Altair
B. Antares
C. Deneb
D. Vega

Mars's largest moon, Phobos, appears over the limb of the planet in this image by the Mars Express *orbiter. Credit: ESA/DLR/FU Berlin (G. Neukum)*

Which of the following is a terrestrial planet?

A. Jupiter
B. Mars
C. Saturn
D. Uranus

Mars's moon Phobos rises in the west and sets in the east. Roughly how long does it take to cross the Martian sky?

A. 1 hour and 37 minutes
B. 2 hours and 24 minutes
C. 4 hours and 15 minutes
D. 6 hours and 41 minutes

July 9th

This image of Jupiter's Great Red Spot was taken by the Voyager 2 *space probe at a distance of 5.99 million km (3.72 million miles). Credit: NASA/JPL.*

In what year did the *Voyager 2* space probe make its closest approach to Jupiter?

A. 1978
B. 1979
C. 1980
D. 1981

Which planet is the only one to turn in a clockwise motion on its axis?

A. Earth
B. Mars
C. Saturn
D. Venus

Saturn's tiny moon Pandora was one of a number discovered in images from the Voyager *space probes. Credit: NASA/JPL-Caltech.*

How are the planets Uranus and Neptune best classified?

A. Exoplanets
B. Gas giants
C. Ice giants
D. Rocky giants

How many moons of Saturn were discovered from images taken by the two *Voyager* space probes?

A. One
B. Three
C. Five
D. Seven

July 11th

Sunset, as seen from Gale Crater on the surface of Mars. Credit: Mars Science Laboratory (MSL), Curiosity, MastCam, NASA/JPL-Caltech/MSSS/Texas A&M Univ.

Which planet has the largest confirmed crater in the solar system?

A. Earth
B. Mars
C. Mercury
D. Venus

When did the earliest forms of life first appear on Earth?

A. 1.3 billion years ago
B. 2.9 billion years ago
C. 3.2 billion years ago
D. 4.1 billion years ago

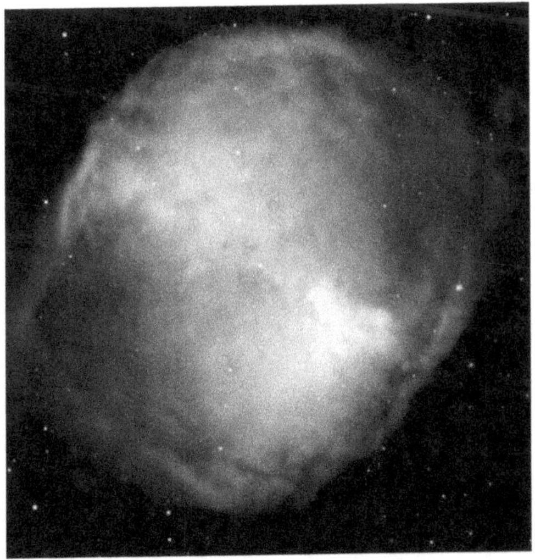

The Dumbbell Nebula, a favorite planetary nebula for many astronomers. Credit: ESO

Who discovered Messier 27, the Dumbbell Nebula, on this date in 1764?

A. William Herschel
B. Jean-Jacques d'Ortous de Mairan
C. Pierre Méchain
D. Charles Messier

Which moon of Neptune makes up more than 99.5% of the total mass of material in orbit about the planet?

A. Triton
B. Nereid
C. Larissa
D. Despina

July 13th

Jupiter's Great Red Spot, seen in unprecedented detail by the Juno *probe in July, 2017. Credit: NASA / SwRI / MSSS / Gerald Eichstädt / Seán Doran.*

What type of object in the solar system can have an atmosphere larger than the Sun?

A. Asteroids
B. Comets
C. Planets
D. Nothing

How many Earths could fit within Jupiter's Great Red Spot?

A. 2 or 3
B. 4 or 5
C. 6 or 7
D. More than 7

An artist's impression of the New Horizons *probe at Pluto. Credit: NASA/Johns Hopkins University Applied Physics Laboratory/Southwest Research Institute*

Which probe was the first to successfully reach Mars and send back close-up images of the planet in 1965?

A. Mariner 4
B. Mariner 6
C. Mariner 9
D. Mariner 10

In what year did the *New Horizons* space probe become the first to flyby the dwarf planet Pluto?

A. 2013
B. 2014
C. 2015
D. 2016

July 15th

This image shows the Earth and Neptune to scale. Credit: NASA.

Which probe became the first to enter the asteroid belt on this date in 1972?

A. *Pioneer 9*
B. *Pioneer 10*
C. *Pioneer 11*
D. *Voyager 1*

What is Neptune's diameter in comparison to the Earth's?

A. 2x Earth
B. 3x Earth
C. 4x Earth
D. 5x Earth

The first image of the asteroid Vesta returned from orbit. Credit: NASA/JPL-Caltech/UCLA/MPS/DLR/IDA.

In which year did the first fragments of Comet Shoemaker-Levy 9 impact Jupiter on this date?

A. 1991
B. 1992
C. 1993
D. 1994

Which space probe became the first to orbit the asteroid Vesta on this date in 2011?

A. *Dawn*
B. *Deep Space 1*
C. *Hyabusa*
D. *NEAR Shoemaker*

July 17th

The star Vega is the brightest star in the summer constellation of Lyra, the Lyre. Credit: Stephen Rahn.

What was the first star to be photographed on this date in 1850?

A. Arcturus
B. Capella
C. Sirius
D. Vega

If Venus is at greatest western elongation, when it is visible?

A. Early evening
B. Throughout the entire night
C. Early morning
D. It's not visible at all

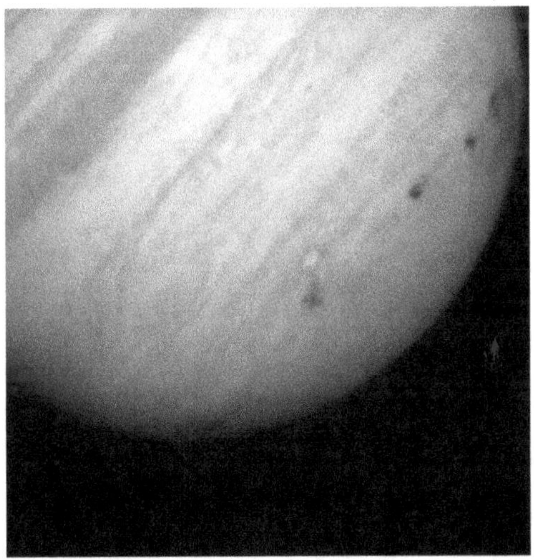

The multiple impact sites of Comet Shoemaker-Levy 9 in the atmosphere of Jupiter. Credit: Hubble Space Telescope Comet Team and NASA.

On this date in 1994 the largest fragment of comet Shoemaker-Levy 9 impacted with the gas giant Jupiter. Roughly how much energy was released as a result?

A. 200x the world's nuclear arsenal
B. 400x the world's nuclear arsenal
C. 600x the world's nuclear arsenal
D. 800x the world's nuclear arsenal

What gas makes up 95% of the Martian atmosphere?

A. Methane
B. Carbon dioxide
C. Nitrogen
D. Oxygen

July 19th

This image from Cassini *shows three moons of Saturn. Titan (right), Rhea (upper left), and Mimas (near the bottom.) Credit: NASA/JPL-Caltech/Space Science Institute.*

How long does it take Mars to complete one orbit of the Sun?

A. 1.9 years
B. 2.1 years
C. 2.3 years
D. 2.5 years

How many moons of Saturn were discovered before the use of long exposure astrophotography?

A. Eight
B. Ten
C. Twelve
D. Fourteen

July 20th

Astronaut Neil Armstrong stands close to the United States flag and the lunar module Eagle. *Credit: NASA.*

When *Apollo 11* landed on the Moon, which “sea” did they land in?

A. Sea of Clouds
B. Sea of Crises
C. Sea of Serenity
D. Sea of Tranquility

On this date in 1976, which space probe was the first to successfully return images from the surface of Mars?

A. *Curiosity*
B. *Pathfinder*
C. *Viking 1*
D. *Voyager 1*

July 21st

The crater Tycho is a prominent feature in the southern hemisphere of the Moon. Credit: NASA, ESA, D. Ehrenreich.

What gas makes up approximately three quarters of the Sun's mass?

A. Helium
B. Hydrogen
C. Methane
D. Nitrogen

Which lunar crater has a bright network of rays that can be easily seen at full Moon?

A. Clavius
B. Picard
C. Theophilus
D. Tycho

July 22nd

The crater Kuiper can be found on the surface of Mercury. Credit: NASA/Johns Hopkins University Applied Physics Laboratory/Carnegie Institution of Washington.

Which of the following *do not* have features on Mercury named for them?

A. Actors
B. Musicians
C. Painters
D. Writers

How much more massive would Jupiter need to be in order to ignite and become a star?

A. 25x more massive
B. 50x more massive
C. 75x more massive
D. 100x more massive

July 23rd

Future colonists explore the surface of Mars. In the background is Pavonis Mons, a volcano on the planet's equator. Credit: NASA/Pat Rawlings.

What's the average distance from the Sun to Neptune? (1 AU = the distance from the Earth to the Sun.)

A. 25 AU
B. 30 AU
C. 35 AU
D. 40 AU

What's the maximum temperature you might experience during summer at the Martian equator?

A. 15° C (59° F)
B. 25° C (77° F)
C. 35° C (95° F)
D. 45° C (113° F)

This mosaic of images from the Lunar Reconnaissance Observer *centers on Oceanus Procellarum, the Ocean of Storms. Credit: NASA.*

Roughly when was the most recent reversal of the Earth's magnetic field?

A. 200,000 years ago
B. 700,000 years ago
C. 1.2 million years ago
D. 1.7 million years ago

What percentage of the Moon's visible near side is covered by the lunar "seas"?

A. 31%
B. 46%
C. 61%
D. 76%

July 25th

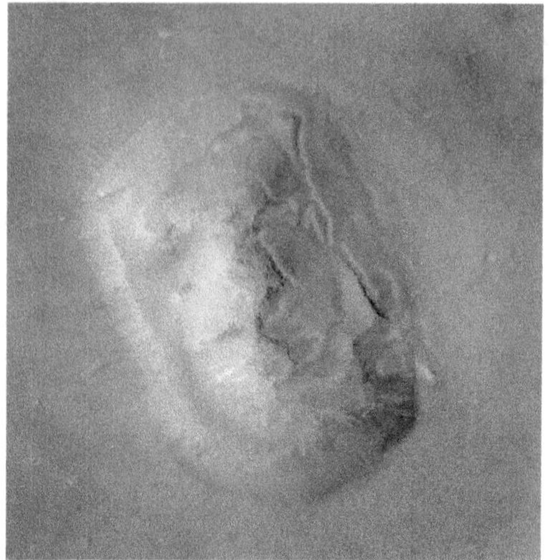

In high resolution images, the "face on Mars" all but disappears. Credit: NASA/JPL/Malin Space Science Systems.

'The teapot' is a familiar pattern of stars within which constellation?

A. Ursa Major
B. Orion
C. Sagittarius
D. Scorpius

The infamous "face on Mars" photo was taken by the space probe *Viking 1* on this date in 1976. In which area of the planet can it be found?

A. Acidalia Planitia
B. Arabia Terra
C. Cydonia
D. Syrtis Major

July 26th

This artist's impression shows what Uranus might look like from its moon Miranda. Credit: JCPag2012 (from Wikimedia).

What's the smallest of the five main moons of Uranus?

A. Ariel
B. Miranda
C. Oberon
D. Titania

What name did Chinese astronomers give to Venus when it appeared in the evening sky?

A. The Excellent West One
B. The Great White Star
C. The Guiding Light of the Fading Sun
D. The Shining Beacon of the Night

July 27th

The four largest moons of Jupiter can easily be seen with a small telescope or binoculars on either side of the planet. Credit: Jan Sandberg.

How long does it take Saturn to complete one orbit of the Sun?

A. 28.5 years
B. 29.0 years
C. 29.5 years
D. 30.0 years

How many moons of Jupiter were discovered by the *Voyager* space probes in 1979?

A. Three
B. Five
C. Seven
D. Nine

July 28th

Comets have been studied and recorded for millennia. This image shows the great comet of 1811. Credit: Public domain.

Comet Halley makes its next closest approach to the Sun on this date – but in which year?

A. 2055
B. 2061
C. 2066
D. 2071

Which of the following is not a superior planet?

A. Jupiter
B. Mars
C. Saturn
D. Venus

July 29th

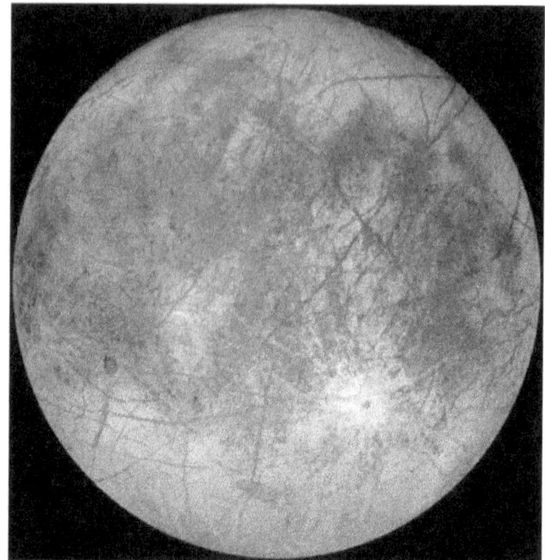

Jupiter's moon Europa is thought to have a liquid ocean beneath its frozen surface. Credit: NASA/JPL/DLR.

How long is a day on Pluto?

A. 1 day and 21 hours
B. 3 days and 18 hours
C. 6 days and 9 hours
D. 8 days and 6 hours

Which moon has been compared to a primordial Earth and may have complex organic chemistry?

A. Charon (Pluto)
B. Enceladus (Saturn)
C. Europa (Jupiter)
D. Titan (Saturn)

The star cluster Messier 7 can be found close to the tail of Scorpius, the Scorpion. Credit: ESO.

Which planet was considered as an additional target for the space probe *Cassini* following the successful completion of its mission at Saturn?

A. Mars
B. Jupiter
C. Neptune
D. Uranus

Which astronomer has the open star cluster Messier 7 named after him?

A. William Herschel
B. Ptolemy
C. Galileo
D. Copernicus

July 31st

The first quarter Moon, June 14th, 2005. Credit: Author.

What was the first moon of Pluto to be discovered?

A. Charon
B. Hydra
C. Nix
D. Styx

When can the first quarter Moon be seen over the southern horizon?

A. Midday
B. Midnight
C. Sunrise
D. Sunset

August

August 1st

This artist's impression depicts the Great Comet of 1861. Credit: E. Weiß.

Which meteor shower is thought to originate from Comet Encke?

A. Geminids
B. Leonids
C. Perseids
D. Taurids

What's the meaning of the original Greek word for comet?

A. Long haired star
B. Misty tailed star
C. Long cloudy star
D. Misty guest star

August 2nd

What might Mars look like from the surface of its moon Phobos? This artist's impression attempts to give us some idea. Credit: NASA

How big does the moon Phobos appear in the Martian sky?

A. About 1/4 the size of our Moon
B. About 1/3 the size of our Moon
C. About 1/2 the size of our Moon
D. About the size of our Moon

How are the planets Jupiter and Saturn best classified?

A. Gas giants
B. Ice giants
C. Rocky Giants
D. Super Earths

August 3rd

Saturn's moon Epimetheus, imaged by the Cassini *space probe. Credit: NASA/JPL-Caltech/Space Science Institute.*

Which space probe, launched on this date in 2004, became only the second to visit the planet Mercury?

A. *Cassini*
B. *Galileo*
C. *Mariner 10*
D. *MESSENGER*

What was the first moon of Saturn to be discovered?

A. Epimetheus
B. Mimas
C. Rhea
D. Titan

The star Fomalhaut (Alpha Piscis Austrini) is an A-type star. Credit: NASA, ESA, and the Digitized Sky Survey 2. Acknowledgment: Davide De Martin (ESA/Hubble)

What gas or element gives Uranus its blue-green color?

A. Hydrogen
B. Methane
C. Nitrogen
D. Oxygen

What type of star is the Sun?

A. K-type
B. A-type
C. O-type
D. G-type

August 5th

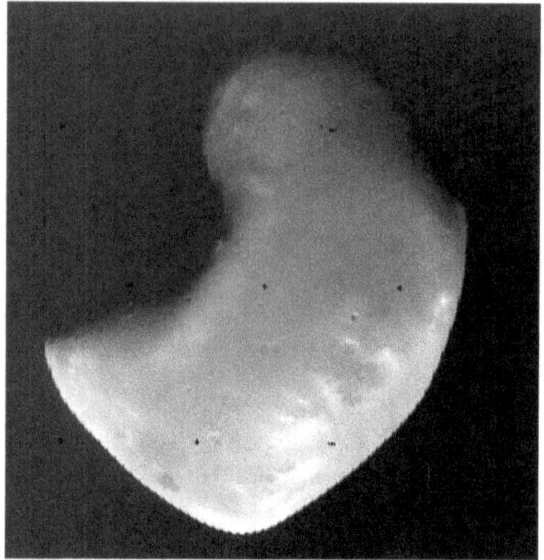

This image of Deimos was taken by the Viking 1 *space probe in 1977. Credit: NASA*

What's the average distance from Mercury to the Sun? (1 AU = the average distance from the Earth to the Sun.)

A. 0.39 AU
B. 0.48 AU
C. 0.52 AU
D. 0.64 AU

How many features on Mars's moon Deimos have been named?

A. Two
B. Twenty
C. Two hundred
D. Two thousand

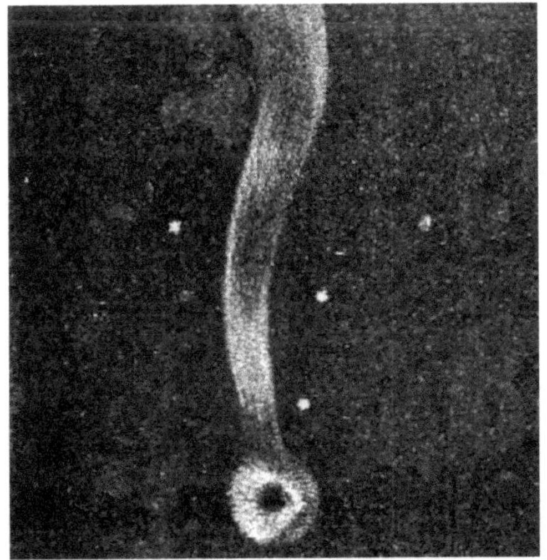

An observation of Halley's Comet as it appeared in 1682. Credit: Artist unknown.

Which civilization recorded observations of Comet Halley on this date in 87 B.C.?

A. Babylonian
B. Egyptian
C. Greek
D. Roman

On this date in 1996, NASA held a press conference regarding a potentially exciting discovery related to the ALH 84001 meteorite from Mars. What was the discovery?

A. An unusually high quantity of iron
B. Possible microscopic bacteria fossils
C. Higher than expected oxygen molecules
D. Evidence of water erosion

August 7th

The crescent Moon, photographed by a crewmember of Expedition 24 *to the* International Space Station. *Credit: NASA.*

What is the average distance from the Earth to the Moon?

A. 302,900 km (188,200 miles)
B. 343,800 km (213,600 miles)
C. 384,500 km (238,900 miles)
D. 423,900 km (263,400 miles)

Which of the following was one of the runner-up names for Pluto?

A. Minerva
B. Hades
C. Orpheus
D. Styx

August 8th

The crescent Moon with the darkened surface illuminated. Venus is to the upper left while Jupiter appears to the upper right in this image. Credit: Sachin Nigam.

When the Moon is a crescent the darkened surface is often visible. What's the name given to this effect?

A. Earthshine
B. An eclipse
C. Moonshine
D. Terralight

Which of the following stars was thought to be the north pole star when the pyramids were built?

A. Arcturus
B. Polaris
C. Thuban
D. Vega

August 9th

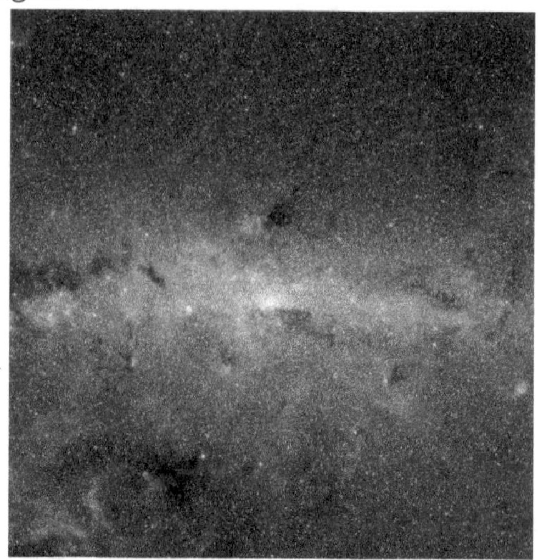

This image shows the heart of our own galaxy in infrared light. Credit: NASA/JPL-Caltech/S. Stolovy (SSC/Caltech)

What's the average distance from the Sun to Saturn? (1 AU = the average distance from the Earth to the Sun.)

A. 6.8 AU
B. 7.2 AU
C. 8.4 AU
D. 9.6 AU

Roughly how far from the center of the galaxy is the Sun?

A. 22,000-25,000 LY
B. 25,000-28,000 LY
C. 28,000-31,000 LY
D. 31,000-34,000 LY

August 10th

The Royal Observatory in Greenwich, England, July 2009. Credit: Mike Peel (www.mikepeel.net).

Which object, first observed by Charles Messier on this date in 1773, was not included in his catalog until after a sketch of his observation was found?

A. M40
B. M73
C. M102
D. M110

The foundation stone of the Royal Observatory in Greenwich, England was laid on this date in which year?

A. 1625
B. 1675
C. 1725
D. 1775

August 11th

The total solar eclipse on August 11th 1999 was the last of the 20th century. Credit: MKcray (from Wikimedia.)

The last total solar eclipse of the 20th century occurred on this date in 1999. Which of the following cities enjoyed over two minutes of totality?

A. London
B. Paris
C. Stuttgart
D. Vienna

Which comet may be responsible for meteor showers affecting the planet Mercury?

A. Encke
B. Hale-Bopp
C. Halley
D. Swift-Tuttle

August 12th

An artist's impression of the Perseid meteor shower of 2006. Credit: Olga Berrios.

The Perseids are at their best at around this date. How many shooting stars may be seen every hour under ideal conditions?

A. 50
B. 75
C. 100
D. 150

In 1987 the first cometary system was discovered surround which star?

A. Alpha Centauri
B. Beta Pictoris
C. Tau Ceti
D. Vega

August 13th

Our galaxy appears edge-on to us in the night sky and is one of about two trillion that are thought to exist in the universe. Credit: ESO/S. Brunier

The winds on Saturn can reach up to what speed?

A. 900 kph (559 mph)
B. 1,200 kph (746 mph)
C. 1,500 kph (932 mph)
D. 1,800 kph (1,118 mph)

What is the name given to the galaxy in which we live?

A. The Andromeda Galaxy
B. The Home Galaxy
C. The Milky Way
D. The Orion Arm

This crater, in Thaumasia Planum on Mars, is 50 km in diameter. Credit: ESA/DLR/FU Berlin (G. Neukum)

How long is a day on Saturn?

A. 7 hours, 23 minutes
B. 8 hours, 11 minutes
C. 9 hours, 56 minutes
D. 10 hours, 33 minutes

Roughly how many craters are on the surface of Mars?

A. 43,000
B. 47,000
C. 51,000
D. 55,000

August 15th

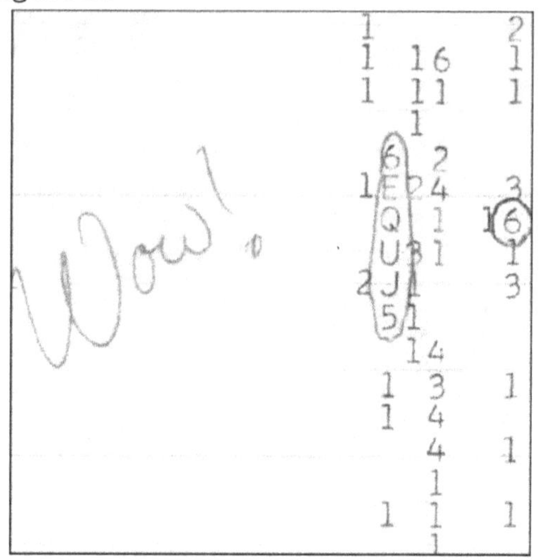

The "Wow!" signal. Credit: Big Ear Radio Observatory and North American AstroPhysical Observatory (NAAPO).

The "Wow!" signal is detected by astronomers using the Big Ear radio telescope on this date in 1977. What was the "Wow!" signal?

A. The first gamma ray burst detected
B. X-rays providing the first evidence of black holes
C. A one-off signal from a possible alien civilization
D. The detection of radiation from a solar flare

Who discovered the two moons of Mars five nights apart in August 1877?

A. Asaph Hall
B. Percival Lowell
C. Giovanni Schiaparelli
D. Louis Thollon

August 16th

The double star Albireo has gold and blue components and is split with a small telescope. Credit: Henryk Kowalewski

In what year did a powerful solar flare affect microchips and cause the Tokyo stock exchange to temporarily suspend trading?

A. 1984
B. 1989
C. 1994
D. 1999

In which constellation is Albireo, arguably the finest double star in the northern hemisphere?

A. Aries
B. Cygnus
C. Orion
D. Sagittarius

August 17th

An artist's impression of the Viking *orbiter releasing the lander while in orbit about Mars. Credit: Don Davis.*

On this date in 1958, an unmanned space probe was launched in the first attempt to reach the Moon. What was it called?

A. *Ranger*
B. *Luna*
C. *Pioneer*
D. *Zond*

NASA deactivated the *Viking 1* orbiter on this date in 1980. It had been orbiting Mars for how many years?

A. Two
B. Four
C. Six
D. Eight

August 18th

Messier 17 in the constellation Sagittarius is a relatively easy target for small telescopes. Credit: Author.

Who made the first (false) report of a satellite of Venus on this date in 1686?

A. Giovanni Cassini
B. Galileo Galilei
C. Edmond Halley
D. Christiaan Huygens

Which of the following is *not* a name for the Messier 17 nebula in the constellation of Sagittarius?

A. The Horseshoe Nebula
B. The Swan Nebula
C. The Omega Nebula
D. The Lagoon Nebula

August 19th

This artist's impression depicts distant Quaoar. Discovered in 2002, it's roughly 43 times further from the Sun than the Earth. Credit: NASA and G. Bacon (STScI)

Which dwarf planet is the most massive?

A. Ceres
B. Eris
C. Pluto
D. Quaoar

What did Galileo mistake the rings of Saturn for when he first observed them with his primitive telescope in 1610?

A. He thought the planet was misshapen
B. He thought his telescope was broken
C. Two moons on either side of the planet
D. Two stars on either side of the planet

August 20th

The Sun and Neptune can be seen rising over the frozen surface of Triton in this artist's impression. Credit: ESO/L. Calçada.

What was the first moon of Neptune to be discovered?

A. Larissa
B. Naiad
C. Nereid
D. Triton

Which space probe was launched on this date in 1977?

A. *Cassini*
B. *Galileo*
C. *Viking 1*
D. *Voyager 2*

August 21st

Alpha Delphini (center) is also known by the name Sualocin and is an easy binocular double. Credit: Roberto Mura

What's the daytime surface temperature on Mercury?

A. 204° C (400° F)
B. 316° C (600° F)
C. 427° C (800° F)
D. 538° C (1,000° F)

Which of the following is the only constellation to contain stars named after an astronomer?

A. Cygnus
B. Delphinus
C. Orion
D. Ursa Major

August 22nd

The Mars Observer *space probe was designed to study the surface, atmosphere, climate and magnetic field of the red planet. Credit: NASA/JPL.*

Which of the following constellations represents a fox?

A. Aquila
B. Cygnus
C. Pyxis
D. Vulpecula

What happened to the *Mars Observer* space probe on this date in 1993?

A. It arrived at Mars
B. It was launched from Cape Canaveral
C. It lost contact with ground control
D. It passed by the Earth

August 23rd

This is the first view of the Earth from lunar orbit, taken by the Lunar Orbiter 1 *space probe. The image was reprocessed and re-released in 2008. Credit: NASA/Lunar Orbiter Image Recovery Project.*

Who made the first telescopic observation of Mercury?

A. Christiaan Huygens
B. Edmond Halley
C. Galileo Galilei
D. William Herschel

In what year on this date did the *Lunar Orbiter 1* probe take the first picture of the Earth from lunar orbit?

A. 1965
B. 1966
C. 1967
D. 1968

This artist's impression shows what the surface of the planet orbiting Proxima Centauri might be like. Credit: ESO/M. Kornmesser.

In what year did the International Astronomical Union vote to redefine and demote Pluto as a dwarf planet?

A. 2003
B. 2004
C. 2005
D. 2006

In what year was the existence of a planet orbiting Proxima Centauri confirmed?

A. 2014
B. 2015
C. 2016
D. 2017

August 25th

Neptune, imaged by the Voyager 2 *space probe. Credit: NASA/JPL.*

On this date in 1835, the *New York Sun* publishes a fake news story about the Moon which was later called the Great Moon Hoax. What was the story?

A. That the Moon was going to explode
B. That the Moon would dramatically shrink
C. That the Moon had life
D. That the Moon was hollow

In what year on this date did *Voyager 2* become the first space probe to pass the planet Neptune?

A. 1988
B. 1989
C. 1990
D. 1991

August 26th

Which probe (top right) made a fly-by of Saturn on this date in 1981? Credit: Don Davis.

Which probe made its closest approach to Saturn on this date in 1981?

A. *Pioneer 11*
B. *Voyager 1*
C. *Voyager 2*
D. *Cassini*

When the space probe *Voyager 2* passed the planet Neptune it sent back images of an atmospheric storm similar to Jupiter's Great Red Spot. What was it called?

A. The Great Blue Spot
B. Big Blue
C. Great Dark Spot
D. The Scooter

August 27th

The total lunar eclipse of April 2014, as seen rising over the mountains. Credit Denali National Park and Preserve.

In 413 B.C., what phenomenon caused the commander of the Athenian army to delay evacuating his troops and thereby allowed them to be defeated by the Spartans?

A. A total solar eclipse
B. A total lunar eclipse
C. The appearance of a comet
D. A close conjunction of Mars and Jupiter

What event in 1883 caused the Moon to literally appear blue?

A. Unusually high oxygen levels in the atmosphere
B. Dust and ash from forest fires in Canada
C. Unusually high rain fall across most of the planet
D. Dust and ash from the eruption of Krakatoa

August 28th

Enceladus, as imaged by the Voyager 2 *space probe. Credit: NASA/JPL.*

Who discovered Saturn's moon Enceladus on this date in 1789?

A. Giovanni Cassini
B. William Herschel
C. Jean-Dominique Maraldi
D. Charles Messier

What unexpected phenomenon was widely witnessed across Europe, Japan and the United States in 1859?

A. The northern lights
B. A meteor shower
C. A comet
D. Ice crystals forming a halo around the Moon

August 29th

Mars has been known as the "red planet" for centuries. This image from the Mars Pathfinder *landing site shows the Twin Peaks in the distance. Credit: NASA/JPL.*

How long did signals from *Voyager 2* take to reach the Earth from Neptune?

A. About an hour
B. About 4 hours
C. About 12 hours
D. About a day

What gives the surface of Mars its reddish-orange appearance?

A. Carbon dioxide
B. Copper deposits in the soil
C. Iron oxide
D. Micro-organisms

Triton is Neptune's largest moon and the first to be discovered. This image was taken by Voyager 2 *during its flyby in August, 1989. Credit: NASA/JPL/USGS.*

How many new moons of Neptune did *Voyager 2* discover during its encounter with the planet in 1989?

A. Two
B. Four
C. Six
D. Eight

Which planet is often reported as a UFO when inexperienced observers see it low over the horizon?

A. Mercury
B. Venus
C. Mars
D. Jupiter

August 31st

The unmanned MESSENGER *space probe was the second to visit Mercury. Credit: NASA/JHUAPL.*

What unusual feature did the *Voyager 2* space probe discover on Neptune's moon Triton in 1989?

A. A crater the size of Texas
B. Nitrogen gas geysers
C. Liquid oxygen
D. Frozen subsurface water

Which unmanned space probe was launched on this date in 2004 on a mission to Mercury?

A. Cassini
B. Hermes
C. Mariner 10
D. MESSENGER

September

September 1st

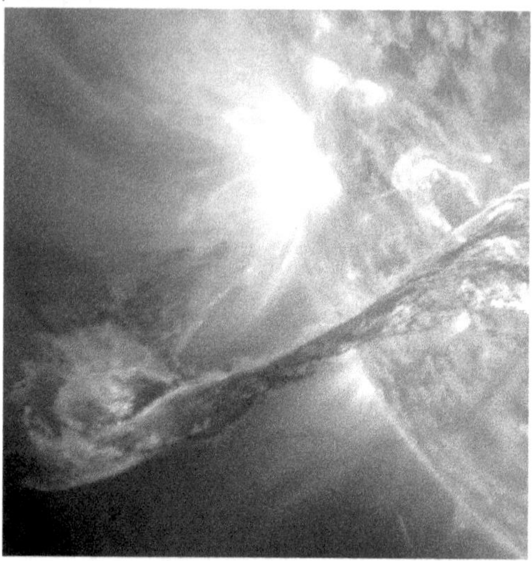

This Coronal Mass Ejection (CME) occurred on August 31st, 2012 and caused aurorae to be seen on September 3rd. Credit: NASA Goddard Space Flight Center

In what year was the first solar flare observed? (It caused widespread disruption across the U.S. on September 2nd.)

A. 1844
B. 1849
C. 1854
D. 1859

Which space probe was the first to visit the planet Saturn on this date in 1979?

A. *Pioneer 10*
B. *Pioneer 11*
C. *Voyager 1*
D. *Voyager 2*

The North American Nebula can be glimpsed under very dark skies with binoculars. Credit: Miodrag Sekulic

What's the average distance between Mars and the Sun? (1 AU = the average distance between the Earth and the Sun)

A. 1.52 AU
B. 1.78 AU
C. 1.91 AU
D. 2.17 AU

Which of the following is not found in the constellation of Cygnus?

A. Albireo
B. The Blinking Planetary
C. The North American Nebula
D. The Wild Duck Cluster

September 3rd

Comet 17P/Holmes was unusual in that it had no tail and glowed a distinctive greenish hue. Credit: Gil-Estel (from Wikimedia).

Jupiter was named after a Roman god but who was his Greek equivalent?

A. Apollo
B. Orpheus
C. Poseidon
D. Zeus

What name is given to the gas and dust that forms an atmosphere around the nucleus of a comet?

A. The atmosphere
B. The coma
C. The halo
D. The tail

September 4th

Could Phobos disintegrate and form a ring about Mars?
Credit: Purdue University Envision Center.

What is one predicted fate for Mars's moon Phobos?

A. It will burn up in the atmosphere of Mars
B. It will collide with Deimos, the other moon
C. It will be destroyed and form a ring around Mars
D. It will escape the gravity of Mars and drift free

What was the first planet discovered by mathematical calculation?

A. Saturn
B. Uranus
C. Neptune
D. Pluto

September 5th

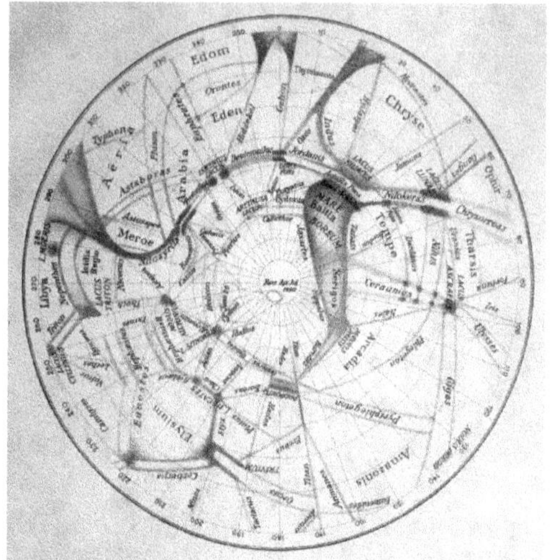

A map of Mars's northern hemisphere produced by Schiaparelli in 1890. Credit: Giovanni Schiaparelli.

In what year did Italian astronomer Giovanni Schiaparelli first observe *canali* on the surface of Mars, leading some to speculate on the existence of intelligent Martian life?

A. 1877
B. 1882
C. 1887
D. 1892

In what year did Pluto reach its minimum distance from the Sun for the first and only time in the 20th century?

A. 1974
B. 1979
C. 1984
D. 1989

September 6th

The Coathanger asterism, in the summer constellation of Vulpecula, the Fox, is easily seen in binoculars. Credit: Petr Novák (http://astrofotky.cz/~karneades)

What is meant when we say Jupiter is a superior planet?

A. The planet is mainly metals and ores
B. It is larger than the Earth
C. It is never visible at midnight
D. It orbits further from the Sun than the Earth

What is an asterism?

A. A group of less than ten stars
B. A group of stars that forms a pattern
C. A group of stars of the same brightness
D. A group of stars of the same color

September 7th

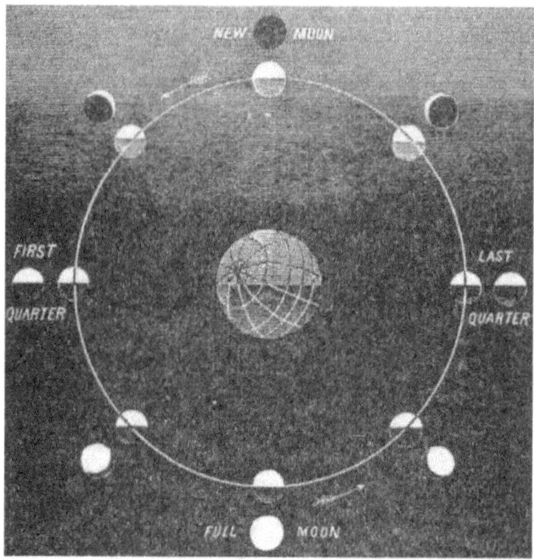

This diagram, from the New Student's Reference Work (1914) shows the phases of the Moon as it orbits the Earth. Sunlight strikes both the Earth and Moon from the top of the image. Credit: Public domain.

In the cycle of lunar phases, which phase comes just before a full Moon?

A. First Quarter
B. Last Quarter
C. Waning Gibbous
D. Waxing Gibbous

What's the nighttime surface temperature on Mercury?

A. -163° C (-261° F)
B. -173° C (-279° F)
C. -183° C (-297° F)
D. -193° C (-315° F)

This view of a crescent Triton (bottom) and Neptune was taken three days after the Voyager 2 *space probe made its closest approach to the planet. Credit NASA/JPL.*

On this date in 2004 the *Genesis* space probe crash landed upon its return to Earth. What cargo was it carrying?

A. Particles from a comet tail
B. Lunar soil samples
C. A sample of solar wind
D. Martian soil samples

How long after the discovery of Neptune was Triton, its largest moon, discovered?

A. 17 days
B. 21 days
C. 25 days
D. 29 days

September 9th

This image of Mars by the Viking 1 *orbiter clearly shows the planet's thin atmosphere. Credit: NASA.*

Which moon of Jupiter was the last to be discovered visually (rather than photographically) in 1892?

A. Amalthea
B. Elara
C. Himalia
D. Pasiphae

What is the atmospheric pressure on Mars compared to the Earth?

A. Less than 1%
B. Between 1% and 5%
C. Between 5% and 10%
D. Between 10% and 15%

September 10th

The Crab Nebula is the remains of a star that exploded over 1,000 years ago. Credit: ESO.

In Roman mythology, what was Mercury's role among the gods?

A. God of the Earth
B. Messenger of the gods
C. Sun god
D. God of war

What is object #1 in Charles Messier's catalog?

A. The Crab Nebula
B. The Keystone Cluster
C. The Orion Nebula
D. The Pleiades

September 11th

A slender crescent Moon can be seen rising through Earth's thick atmosphere. Credit: NASA Earth Observatory

Which moon of Jupiter is the most likely candidate as a home to primitive extra-terrestrial life?

A. Io
B. Europa
C. Ganymede
D. Callisto

What altitude is the official boundary between the Earth's atmosphere and space?

A. 50 km (32 miles)
B. 75 km (47 miles)
C. 100 km (62 miles)
D. 125 km (78 miles)

September 12th

Did astronomers believe they'd found a tenth planet on this date in 1983? Credit: PlanetUser (from Wikimedia)

What did astronomers mistakenly believe they had discovered as they observed Neptune pass in front of a star in 1983?

A. A new moon
B. A new planet
C. Rings around Neptune
D. A supernova

Approximately how long does it take for the Sun to travel one light year through space?

A. 1,190 years
B. 1,373 years
C. 1,591 years
D. 1,728 years

September 13th

This artist's impression shows the formation of the planets from the solar nebula. Credit: NASA/Pat Rawlings.

The true nature of meteors wasn't known until the early 18th century. What did the ancient Greeks think them to be?

A. Atmospheric phenomenon
B. The gods moving across the sky
C. Souls of the recently departed
D. Stars falling from the sky

How long did it take the primordial Earth to form from the solar nebula?

A. 0-10 million years
B. 10-20 million years
C. 20-30 million years
D. 30-40 million years

September 14th

Mimas is a small, heavily cratered Moon of Saturn. Credit: NASA / JPL-Caltech / Space Science Institute.

Which of the following planets is bright enough to cast shadows?

A. Mars
B. Mercury
C. Saturn
D. Venus

What's the name of the large, circular crater on Saturn's moon Mimas that gives the moon its distinctive similarity to the Death Star in *Star Wars*?

A. Cassini
B. Galileo
C. Herschel
D. Huygens

September 15th

The Cassini *space probe burns up in the atmosphere of Saturn at the end of its mission. Credit: Kevin Gill.*

On this date in 2011, the extrasolar planet Kepler-16b is discovered some 200 light years away. What's so special about it?

A. It's first found orbiting a binary star system
B. It's the furthest exoplanet discovered
C. It's the most Earth-like exoplanet discovered
D. It's the first found orbiting a red dwarf star

The unmanned *Cassini* mission to Saturn came to an end on this date in which year?

A. 2014
B. 2015
C. 2016
D. 2017

September 16th

Although rocky, Hyperion's surface looks bizarrely sponge-like. Credit: NASA/JPL/Space Science Institute.

In 2011, the *Cassini* space probe makes a flyby of which Saturnian moon on the 163rd anniversary of its discovery?

A. Enceladus
B. Hyperion
C. Phoebe
D. Titan

On this date in 1835, the *New York Sun* newspaper admits to perpetuating which hoax?

A. Alien life visiting Earth
B. Life on Mars
C. Life on the Moon
D. Life on the Sun

September 17th

William Herschel, discoverer of Uranus, numerous moons and a prolific astronomer. Credit: Public domain.

Which Saturnian moon did William Herschel discover on this date in 1789?

A. Dione
B. Iapetus
C. Mimas
D. Tethys

In 2008 the International Astronomical Union announces the name of the dwarf planet Haumea. Whose team is credited with its controversial discovery?

A. Mike Brown
B. Geoff Marcy
C. José Luis Ortiz Moreno
D. No one's

September 18th

This image of a crescent Earth and Moon together was the first of its kind taken by any spacecraft. Credit: NASA.

Which space probe took the first ever image of the Earth and Moon together on this date in 1977?

A. *Pioneer 10*
B. *Pioneer 11*
C. *Voyager 1*
D. *Voyager 2*

In Roman mythology, what did the god Saturn represent?

A. Agriculture
B. Education
C. Love
D. War

September 19th

This view of Saturn and its rings was taken by the Voyager 1 *space probe, four days after its closest approach to the planet in November 1980. Credit: NASA.*

What rare astronomical event was observed by astronomers in 1170?

A. An impact on the Moon
B. Mars passing in front of Jupiter
C. A supernova
D. Mercury passing in front of Venus

What are the rings of Saturn primarily made of?

A. Dust
B. Rock
C. Methane ice particles
D. Water ice particles

September 20th

Taken on December 7th, 1972 by the crew of Apollo 17, *this famous photograph was nicknamed the Blue Marble and became one of the most downloaded images of all time. Credit: NASA.*

What's the Greek name for the Earth?

A. Gaia
B. Ishtar
C. Sol
D. Terra

Which of the following was a suggested name for Neptune when it was first discovered?

A. Cronus
B. Janus
C. Markab
D. Odysseus

September 21st

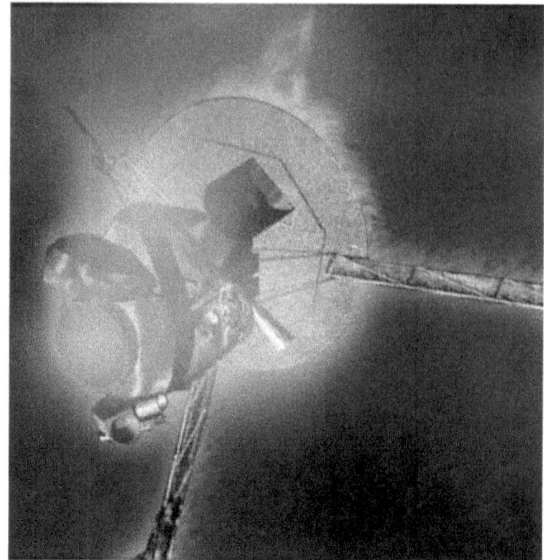

This artist's impression depicts the space probe Galileo *as it burns up in the atmosphere of Jupiter. Credit: NASA*

In what year did NASA end the *Galileo* mission by having the probe burn up in the atmosphere of Jupiter?

A. 2002
B. 2003
C. 2004
D. 2005

When is Mercury or Venus at inferior conjunction?

A. It appears close to the Moon in the sky
B. It appears close to another planet in the sky
C. It lies directly between the Earth and the Sun
D. It's on the opposite side of the Sun

September 22nd

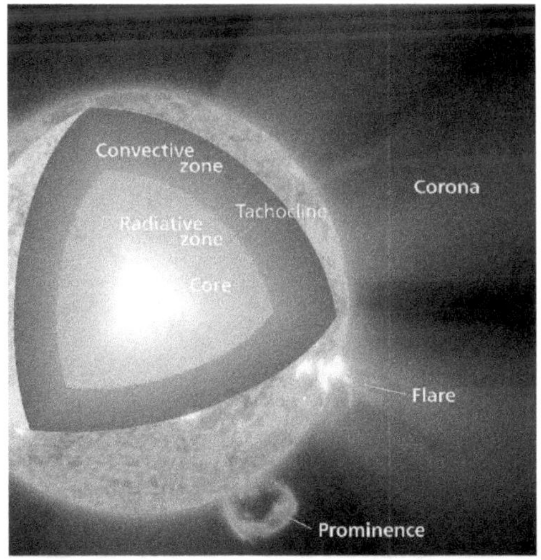

Neutrinos must travel nearly 700,000 km (435,000 miles) from the core to the surface of the Sun. Credit: Kelvinsong (from Wikimedia)

Which planet has a radius one tenth of the Sun's?

A. Saturn
B. Neptune
C. Uranus
D. Jupiter

How long does it take a neutrino to travel from the core to the surface of the Sun?

A. 2.3 seconds
B. 2.3 minutes
C. 2.3 hours
D. 2.3 days

September 23rd

Johann Galle, around 1890, nearly fifty years after his discovery of Neptune. Credit: Olga Radomsky.

On this date in 1846, Johann Galle discovers the planet Neptune after receiving a letter predicting its position from Urbain Le Verrier. For how long did Galle search the skies after receiving the letter?

A. It was discovered that night
B. One week
C. One month
D. One year

What's the size of most meteors?

A. About the size of a grain of sand
B. About the size of a penny
C. About the size of a small stone
D. About the size of a baseball

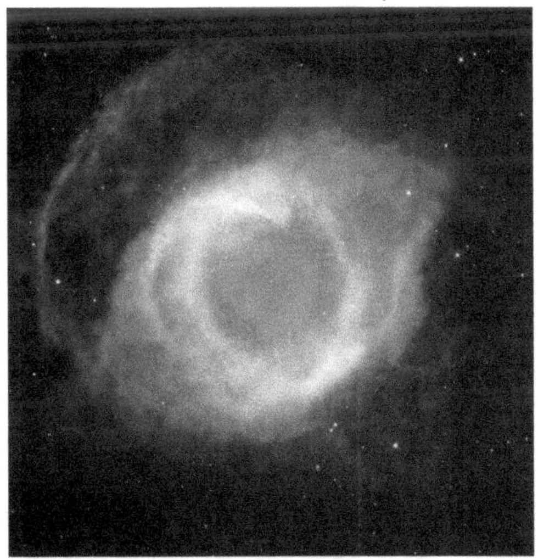

The faint Helix Nebula can be found in the constellation of Aquarius. Credit: NASA, ESA, and C.R. O'Dell (Vanderbilt University)

Which of the following terms is a measure of an object's brightness in the night sky?

A. Albedo
B. Contrast
C. Depth
D. Magnitude

Which of the following autumn constellations does *not* have a direct connection to water?

A. Aquarius
B. Cetus
C. Pegasus
D. Pisces

September 25th

Saturn's moon Rhea appears against a backdrop formed by the giant planet and its rings in this image by the Cassini *space probe. Credit: NASA/JPL/Space Science Institute*

What is thought to have caused the planet Uranus to lie on its side rather than sit upright like the rest of the planets?

A. It was formed that way
B. An Earth-sized proto-planet collided with it
C. The passing of a nearby comet disrupted it
D. Its magnetic poles flipped

Which of the following moons is *not* larger than Mercury?

A. Ganymede (Jupiter)
B. The Moon (Earth)
C. Rhea (Saturn)
D. Titan (Saturn)

September 26th

The Leviathan of Parsonstown had an aperture of 72 inches (6 feet.) Credit: Public Domain.

Who had a telescope that was nicknamed "the Leviathan of Parsonstown" by the locals?

A. Galileo Galilei
B. William Herschel
C. Charles Messier
D. The third Earl of Rosse

On this date in 1950, the full Moon literally appeared blue to observers. What caused this?

A. Smoke from wildfires in Canada
B. An excess of nitrogen in the atmosphere
C. Ash and dust from a volcano
D. High altitude ice particles

September 27th

When Saturn's rings appear edge-on from Earth, they can literally disappear from view. This image was taken with the rings narrowly open. Credit: NASA/ESA/STScI/AURA.

How much more massive is Neptune compared to the Earth?

A. 12x
B. 17x
C. 22x
D. 27x

Roughly how often do Saturn's rings apparently vanish as they appear edge-on to us?

A. Every 5 years
B. Every 10 years
C. Every 15 years
D. Every 20 years

September 28th

Carl Sagan's Cosmos television series has been seen by more than half a billion people in 60 countries. Credit: NASA/JPL.

In what year did Carl Sagan's *Cosmos* television series premiere on PBS in the United States?

A. 1978
B. 1979
C. 1980
D. 1981

What was the first comet to be successfully photographed on this date in 1858?

A. Donati's Comet
B. Encke's Comet
C. Comet Hale-Bopp
D. Halley's Comet

September 29th

Evelyn de Morgan's 1885 painting depicts the Moon goddess bound by the ropes of fate. Credit: Public domain.

How long does it take for Mercury to spin on its axis?

A. 13.3 days
B. 28.9 days
C. 43.2 days
D. 58.7 days

What was the name of the Greek lunar goddess?

A. Diana
B. Minerva
C. Selene
D. Vesta

The very first image of the Orion Nebula, a fifty minute exposure taken with an 11 inch refractor in the late 19th century. Credit: Henry Draper.

Henry Draper took the first successful photograph of a nebula (the Orion Nebula) on this date in which year?

A. 1865
B. 1870
C. 1875
D. 1880

Which planet exhibits the atmospheric phenomenon known as the Ashen Light?

A. Venus
B. Earth
C. Mars
D. Jupiter

October

October 1st

The Cassini Division, a gap that separates the rings of Saturn, can clearly by seen In this sketch of the planet by Christiaan Huygens in 1676. Credit: Public domain.

In what year did Giovanni Cassini first observe the division in Saturn's rings that now bears his name?

A. 1670
B. 1675
C. 1680
D. 1685

Which planet has the smallest axial tilt?

A. Mercury
B. Earth
C. Saturn
D. Neptune

The Voyager *probes visited the outer planets and their moons in the 1970's and 1980's, but which planet was the first to be discovered telescopically? Credit: Don Davis.*

Which planet was not known to the ancient Greeks and only discovered after the invention of the telescope?

A. Mars
B. Jupiter
C. Saturn
D. Uranus

The Sun accounts for how much of the total mass of the solar system?

A. 84.6%
B. 89.8%
C. 94.2%
D. 99.9%

October 3rd

Meteor Crater in Arizona was formed when a 150 foot (45 meter) rock struck the Earth. Credit: Author.

What's the difference between a meteor and a meteorite?

A. Meteors are only seen during the summer
B. Meteorites will hit the ground
C. Meteorites are too faint to be seen
D. Meteors are always linked to a constellation

What chemical gives some meteors a violet color?

A. Calcium
B. Magnesium
C. Potassium
D. Sodium

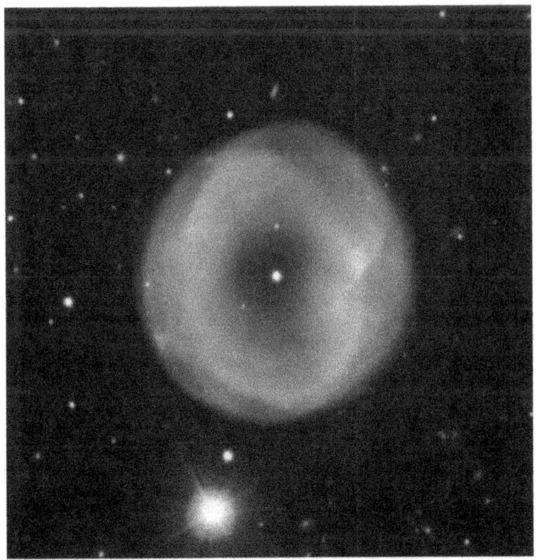

IC 5148, the Spare Tire Nebula, is a planetary nebula 3,000 light years away in the constellation Grus. Credit: ESO.

Which giant planet is the densest?

A. Jupiter
B. Neptune
C. Saturn
D. Uranus

What does the modern constellation Grus represent?

A. A swan
B. A giraffe
C. A crane
D. A bear

October 5th

An artist's impression of Pluto and its moon Charon (foreground) prior to the arrival of the New Horizons *space probe in 2015. Credit: Pat Rawlings/NASA.*

What was the informal nickname given to the yet-to-be discovered Pluto in the early 20th century?

A. Planet X
B. Nemesis
C. Poseidon
D. Vulcan

What part of a ship does the southern constellation Carina represent?

A. The bow
B. The sails
C. The stern
D. The keel

October 6th

This artist's impression shows one of the planets orbiting the star 47 Ursae Majoris. Credit: Debivort (via Wikimedia)

In 1995 the first exoplanet found orbiting a sun-like star is discovered. Which star hosts the planet?

A. 47 Ursa Majoris
B. 51 Pegasi
C. 55 Cancri
D. 61 Virginis

Which planet has a mass of 2 ½ times all the other planets combined?

A. Jupiter
B. Saturn
C. Uranus
D. Neptune

October 7th

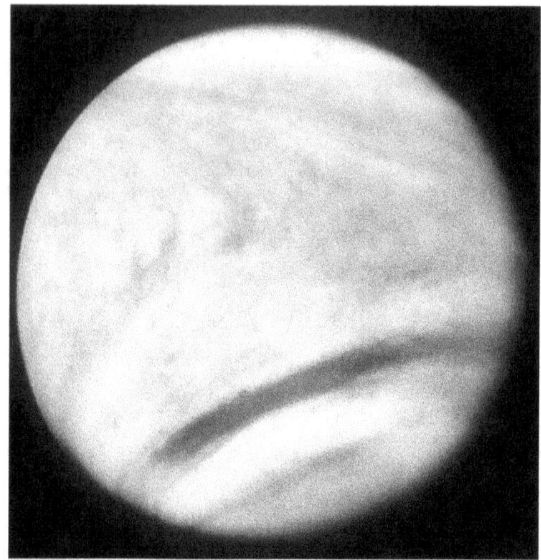

The clouds of Venus cover the planet making its surface impossible to view. Credit: NASA.

What's so special about the tail of a comet?

A. It always points toward the Earth
B. It's too faint to be seen with the naked eye
C. It has micro-biological organisms
D. It nearly always points away from the Sun

Which of the following planets has the thickest atmosphere?

A. Mercury
B. Venus
C. Earth
D. Mars

Pioneer Venus *deployed a number of smaller probes into the atmosphere of Venus. Credit: NASA.*

What happened to the *Pioneer Venus* space probe at the end of its mission in 1992?

A. It flew on to Mercury
B. It was burnt up in Venus's atmosphere
C. It returned to Earth
D. It was simply powered down

What is the original Greek name of the god for whom the planet Uranus is named?

A. Ouranos
B. Cronus
C. Thetis
D. Urania

October 9th

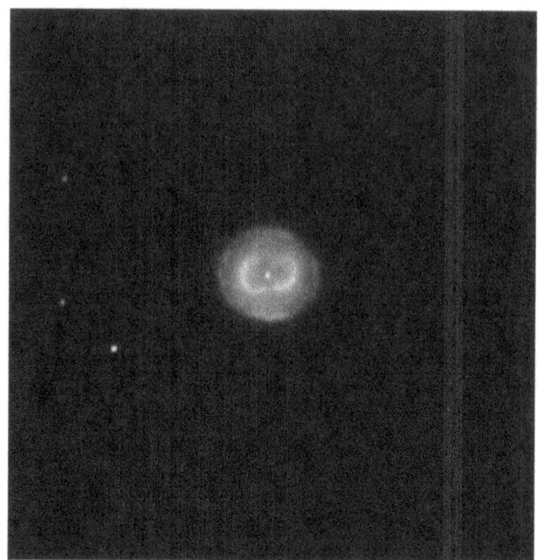

NGC 7662, aka the Blue Snowball Nebula, can be found in the constellation of Andromeda and is a good example of a typical planetary nebula. Credit: Gianluca Pollastri

How many moons does Venus have?

A. None
B. 1
C. 3
D. 5

Why are planetary nebulae so-called?

A. Planets are formed there
B. They're about the same size as a planet
C. They look like planets through a telescope
D. They orbit stars like a planet

October 10th

The Andromeda Galaxy can be seen with just the naked eye under dark skies. Credit: Adam Evans.

In what year did William Lassell discover Triton, Neptune's largest moon?

A. 1841
B. 1846
C. 1851
D. 1856

Which of the following cannot be found in the constellation of Andromeda?

A. Almach
B. The Andromeda Galaxy
C. The Blue Snowball nebula
D. The Double Cluster

October 11th

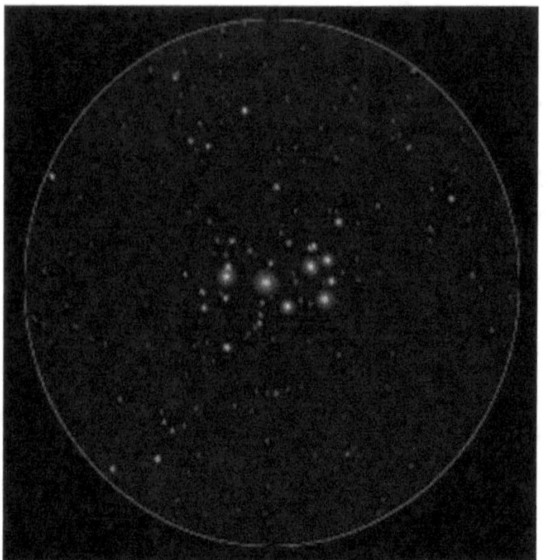

The Pleiades star cluster is considered to be one of the finest sights in the night sky. This image simulates the view through binoculars. Credit: Roberto Mura

Which of the following is not an autumn constellation?

A. Cancer, the Crab
B. Cassiopeia, the Queen
C. Pegasus, the Flying Horse
D. Pisces, the Fishes

The Pleiades star cluster is also known as the Seven Sisters but how many stars can most observers see with the naked eye?

A. Four
B. Six
C. Eight
D. Ten

The MESSENGER *space probe took this image of the Caloris Basin in 2015. Credit: NASA.*

Which of the following is the most distant object that's easily seen with the unaided eye?

A. The Andromeda Galaxy
B. The Hyades star cluster
C. The Orion Nebula
D. The Pleiades star cluster

What's the name of the largest known crater on Mercury?

A. Homer
B. Shakespeare
C. Caloris Basin
D. Rembrandt

October 13th

This image, taken with the Hubble Space Telescope *in 1996, contains several hundred galaxies. Credit: NASA/JPL/STScI Hubble Deep Field Team.*

According to the results of a study released on this date in 2016, how many galaxies are thought to exist in the universe?

A. Two thousand
B. Two million
C. Two billion
D. Two trillion

How long does it take Mercury to rotate once on its axis?

A. 5 hours 54 minutes
B. 2 days 11 hours
C. 5 days 9 hours
D. 59 days

October 14th

This image of Mercury's cratered surface was taken by the MESSENGER *space probe in 2008. Credit: NASA/Johns Hopkins APL/Carnegie Institution of Washington.*

Who recorded an observation of a suspected ring around Neptune in 1846?

A. John Couch Adams
B. Johann Galle
C. William Lassell
D. Urbain Le Verrier

Which of the following planets has a heavily cratered surface that most resembles our Moon?

A. Mars
B. Mercury
C. Jupiter
D. Venus

October 15th

Janus (bottom) and Prometheus (top) are two moonlets of Saturn. Credit: NASA/JPL/Space Science Institute

On this date in 1997 the *Cassini-Huygens* space probe is launched on a mission to Saturn. How long does it take to reach its destination?

A. 5 years
B. 7 years
C. 10 years
D. 12 years

How many moonlets are known to orbit within Saturn's A ring?

A. Between 75 and 100
B. Between 100 and 125
C. Between 125 and 150
D. More than 150

October 16th

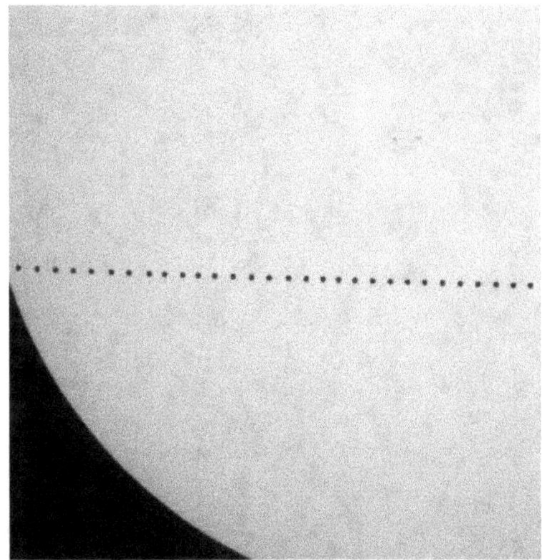

This image shows the transit of Mercury across the Sun's disk on May 9th, 2016. Credit: NASA Goddard Space Flight Center.

On average, how often does a transit of Mercury across the face of the Sun occur?

A. Every 7 years
B. Every 37 years
C. Every 67 years
D. Every 97 years

Which planet has rings that are easily observed through a small telescope?

A. Jupiter
B. Saturn
C. Uranus
D. Neptune

October 17th

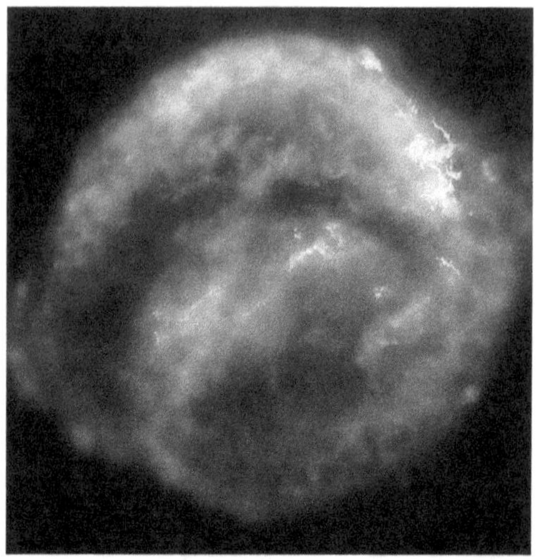

This infrared/optical/X-ray composite image shows the expanding remains of the 1604 supernova. Credit: NASA/ESA/JHU/R.Sankrit & W.Blair

In what constellation was the supernova of 1604 observed by Johannes Kepler?

A. Cassiopeia
B. Cygnus
C. Ophiuchus
D. Pegasus

How much larger is the Sun compared to the Earth?

A. 75xlarger
B. 83x larger
C. 96x larger
D. 109x larger

Like asteroid Chariklo, minor planet Chiron is thought to be surrounded by a faint set of rings. Credit: Celestia.

Which probe became the first to send back data from the atmosphere of another planet (Venus)?

A. *Galileo*
B. *Pioneer 10*
C. *Venera 4*
D. *Viking 1*

The minor planet Chiron was discovered on this date in 1977. Where does it orbit?

A. Between Mars and Jupiter
B. It trails behind Jupiter in its orbit
C. Between Saturn and Uranus
D. Beyond Neptune

October 19th

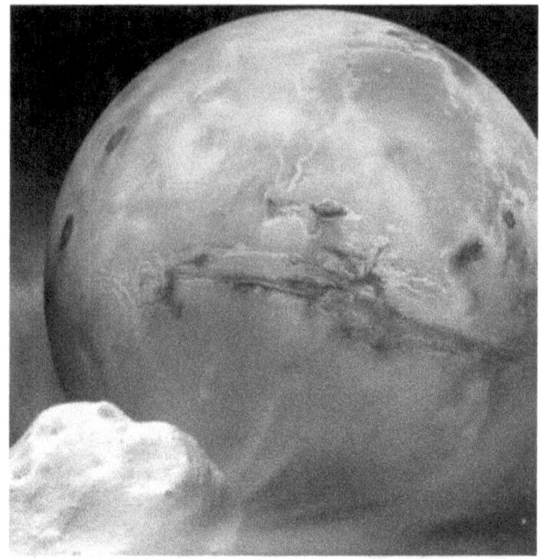

Mars had a close encounter with a comet on this date in 2014. Credit: NASA.

How long does it take Pluto to complete one orbit of the Sun?

A. 246 years
B. 247 years
C. 248 years
D. 249 years

Which comet came within 139,500 km (88,000 miles) of the planet Mars on this date in 2014?

A. Halley
B. Siding-Spring
C. C/2014 PanSTARRS
D. Macholz

October 20th

This computer simulation depicts Neptune rising over the surface of Triton. Credit: NASA

What is Neptune's largest moon?

A. Larissa
B. Nereid
C. Thalassa
D. Triton

Which ancient civilization named the planet Mercury Nabu after their messenger of the gods?

A. The Babylonians
B. The Greeks
C. The Egyptians
D. The Romans

October 21st

Comet West has an orbital period of about half a million years. Credit: J. Linder/ESO

The Orionid meteors peak around this date every year. Which comet is thought to renew the shower as it passes through the solar system?

A. Hale-Bopp
B. Halley
C. Shoemaker-Levy 9
D. Swift-Tuttle

Which comet was visible during daylight hours at this time in 1965?

A. Alcock
B. IRAS-Aracki
C. Ikeya-Seki
D. West

October 22nd

An artist's impression of an exoplanet orbiting 16 Cygni. Credit: NuclearVacuum (from Wikimedia.)

Which space probe was the first to send back an image from the surface of Venus in 1975?

A. *Venera 9*
B. *Venera 10*
C. *Venera 11*
D. *Venera 12*

On this date in 1996, an exoplanet is discovered orbiting the star 16 Cygni. What's special about the discovery?

A. It's the first Jupiter-sized exoplanet found
B. It's the first planet found using direct imaging
C. It's the first planet found orbiting a neutron star
D. It's the first planet found orbiting a triple star

October 23rd

Phobos (top) and Deimos (bottom) are the two rocky moons of Mars. The pair are thought to be captured asteroids. Credit: NASA/JPL/USGS

Mars has two moons, Phobos and Deimos. What do their names mean?

A. Fear and Dread
B. Ferocity and Destruction
C. Victory and Defeat
D. War and Peace

Which planet is the smallest?

A. Earth
B. Venus
C. Mercury
D. Mars

Uranus's moon Ariel, as imaged by the space probe Voyager 2 *on January 24th, 1986. Credit: NASA/JPL.*

Who discovered Uranus's moons Ariel and Umbriel in 1851?

A. John Couch Adams
B. Gerard Kuiper
C. William Lassell
D. Urbain Le Verrier

In what year did a modified German V2 rocket take the first photo of the Earth from space?

A. 1941
B. 1946
C. 1951
D. 1956

October 25th

Saturn's moon Tethys slowly disappears behind its much larger sibling, Titan, as seen in this image by the Cassini *space probe. Credit: NASA/JPL/Space Science Institute*

Saturn's second moon was discovered by Giovanni Cassini on this date in 1671. Which moon was it?

A. Enceladus
B. Iapetus
C. Tethys
D. Titan

Which comet was easily visible to the naked eye around this time in 2007?

A. Holmes
B. Ikeya-Zhang
C. Macholz
D. McNaught

Messier 79 is a globular star cluster containing about 150,000 stars. Credit: NASA/ESA.

Which planet orbits the Sun by rolling on its side rather than moving upright through space?

A. Mercury
B. Neptune
C. Uranus
D. Venus

Pierre Méchain discovers the globular star cluster Messier 79 on this date in 1780. In which constellation can it be found?

A. Canis Major
B. Cassiopeia
C. Gemini
D. Lepus

October 27th

Point a camera toward the north and set a long exposure and you could capture the trails of the stars as they spin about Polaris, the pole star. Credit: Kevin Hadley.

Which of the following constellations is not considered to be circumpolar?

A. Cassiopeia
B. Draco
C. Orion
D. Ursa Major

What is the astronomical symbol for the Earth?

A. A circle with a cross within it
B. A circle with a diagonal line through it
C. A circle with an arrow attached
D. Two circles linked together

October 28th

This image of the Moon rising from Earth orbit was taken in July 1995 by the crew of the space shuttle Discovery. *Credit: NASA*

What is Saturn's mass in comparison with the Earth's?

A. 85x Earth's mass
B. 95x Earth's mass
C. 105x Earth's mass
D. 115x Earth's mass

What's the popular definition of a "blue Moon"?

A. The furthest full Moon from the Earth in a year
B. The nearest full Moon to the Earth in a year
C. The second of two new Moons in a month
D. The second of two full Moons in a month

October 29th

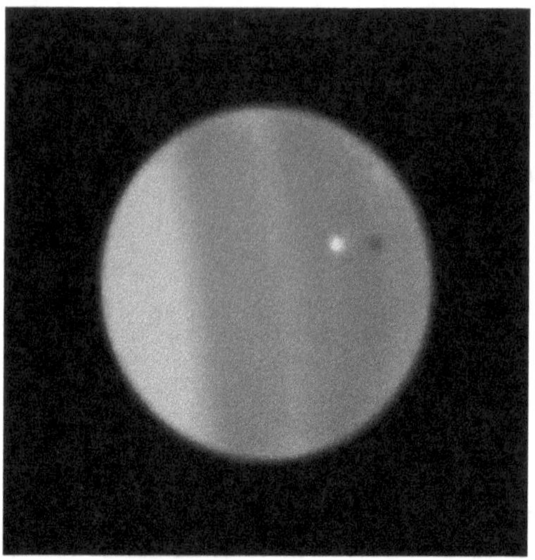

This image of Uranus by the Hubble Space Telescope *shows its moon Ariel (white dot) casting a shadow (black dot) against the planet. Credit: NASA/Space Telescope Science Institute.*

What color does Jupiter appear in our sky?

A. Orange-brown
B. Pinkish
C. White
D. Strong yellow

Uranus's moons Miranda and Ariel are named after characters from which Shakespearean play?

A. *Henry V*
B. *A Midsummer's Night's Dream*
C. *Romeo & Juliet*
D. *The Tempest*

October 30th

This artist's impression depicts the giant planet Jupiter rising over the surface of its moon Europa. Credit: NASA/JPL-Caltech

How strong is Jupiter's magnetic field in comparison to the Earth's?

A. 9x stronger
B. 14x stronger
C. 19x stronger
D. 24x stronger

Besides the Earth, what's the only world in the solar system that has bodies of liquid on its surface?

A. Europa (Jupiter)
B. Mars
C. Titan (Saturn)
D. Triton (Neptune)

October 31st

This orbital view of moonrise over the western United States shows ample evidence of human life – city lights. Credit: NASA

How many constellations are officially recognized by the International Astronomical Union?

A. 55
B. 66
C. 77
D. 88

The population of the Earth hit seven billion on this date in 2011. When will it hit the eight billion mark?

A. 2040
B. 2050
C. 2060
D. 2070

November

November 1st

The Arecibo Radio Telescope has a dish 1,000 feet in diameter and was the largest single aperture telescope in the world until 2016. Credit: JidoBG (from Wikimedia)

In what year did the Arecibo Radio Telescope in Puerto Rico officially open?

A. 1958
B. 1963
C. 1968
D. 1973

In what century was Jupiter's Great Red Spot first observed?

A. 17th century
B. 18th century
C. 19th century
D. 20th century

November 2nd

Mira, a variable star whose brightness changes over a period of almost 498 days, is the bright star in the upper right corner of this image. Credit: DSS 2/ESO.

The star Omicron Ceti is commonly known as Mira. What does its name mean?

A. The Miraculous
B. The Mysterious
C. The Terrible
D. The Wonderful

What is the speed of light?

A. 270,695 km per sec (168,202 miles per second)
B. 286,772 km per sec (178,192 miles per second)
C. 299,792 km per sec (186,282 miles per second)
D. 313,608 km per sec (194,867 miles per second)

November 3rd

The Mariner 10 *space probe passed Venus while en-route to Mercury. Credit: NASA.*

Launched by NASA on this date in 1973, what was the first space probe to encounter the planet Mercury?

A. *Mariner 9*
B. *Mariner 10*
C. *Mercury Planetary Orbiter*
D. *MESSENGER*

Who was the first to correctly identify Saturn's rings as actual rings about the planet?

A. Galileo Galiei
B. Edmond Halley
C. William Herschel
D. Christiaan Huygens

The exoplanet Kepler 22b is just under 2½ times larger than the Earth and orbits within the habitable zone of its parent star. Credit: NASA/Ames/JPL-Caltech

Based on data obtained from the orbiting Kepler observatory, how many Earth-sized planets are thought to be orbiting within habitable zones in our own galaxy?

A. 10 billion
B. 20 billion
C. 30 billion
D. 40 billion

How long does Phobos take to orbit Mars?

A. 7 hours, 39 minutes
B. 8 hours, 7 minutes
C. 9 hours, 13 minutes
D. 10 hours, 28 minutes

November 5th

An artist's impression of the space probe Galileo *passing Io with Jupiter in the background. Credit: NASA/JPL.*

On this date in 2002, the space probe Galileo made its last flyby of a Jovian moon before being de-orbited. Which moon did it encounter?

A. Amalthea
B. Europa
C. Io
D. Thebe

Which country made its first attempt to reach Mars with an unmanned space probe on this date in 2013?

A. Canada
B. China
C. India
D. Japan

November 6th

The constellation Andromeda, as depicted in an astronomical chart dating back to 1825. Credit: Sidney Hall.

Who does the constellation Andromeda represent in Greek mythology?

A. A young girl
B. A servant
C. A princess
D. A queen

Which Saturnian moon is thought to be provide material for the planet's E ring?

A. Enceladus
B. Iapetus
C. Mimas
D. Titan

November 7th

NGC 457 is an open star cluster easily seen with binoculars in the autumn constellation of Cassiopeia, the Queen. Credit: Gabbygall (from Wikimedia.)

Which moon is the largest in the solar system?

A. Ganymede (Jupiter)
B. The Moon (Earth)
C. Titan (Saturn)
D. Triton (Neptune)

Which of the following is *not* a name for the open star cluster NGC 457?

A. The E.T. Cluster
B. The Kachina Doll Cluster
C. The Owl Cluster
D. The Swan Cluster

When we look up at the Milky Way, we're actually looking at our own galaxy edge-on and from within one of it's spiral arms. Credit: Samuel Rubin

What is meant when we say Mercury is an inferior planet?

A. The planet has few metals and ores
B. It's always visible at midnight
C. It orbits closer to the Sun than the Earth
D. It's smaller than the Earth

In which arm of the Milky Way galaxy does our solar system reside?

A. Andromeda
B. Carina-Sagittarius
C. Orion-Cygnus
D. Perseus

November 9th

The astronomer Percival Lowell observers the planet Venus in 1914. Credit: Public domain.

Who initiated and funded the search for a theoretical ninth planet in 1906?

A. George Ellery Hale
B. Edwin Hubble
C. Gerard Kuiper
D. Percival Lowell

What percentage of all life that has ever existed on Earth is now extinct?

A. 54%
B. 69%
C. 84%
D. 99%

November 10th

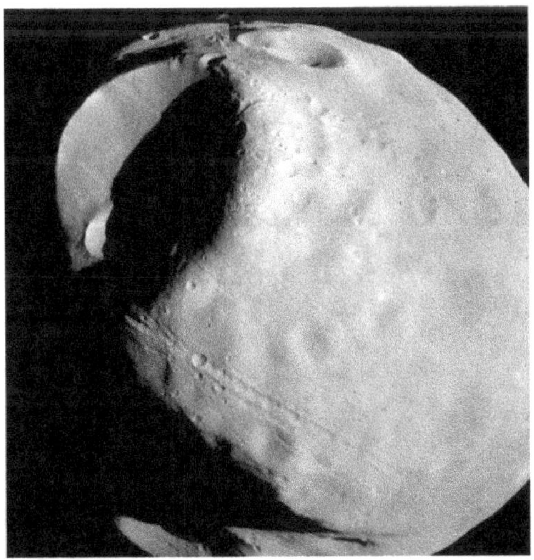

This image of Phobos was created from images taken by Viking 1 *in 1978. Credit: NASA*

What's the name of the 9km wide crater that dominates the surface of Mars' moon Phobos?

A. Hall
B. Lowell
C. Stickney
D. Wells

Which planet has the longest day (243 days)?

A. Mercury
B. Venus
C. Mars
D. Jupiter

November 11th

Viking 1 *dug a number of trenches on the surface of Mars to test the composition of the soil. Credit: NASA/JPL*

The *Viking 1* lander mission came to an end on this date in 1982. It had been operating for six years. Why did the mission end?

A. Lack of funding shut down the Viking program
B. An erroneous command was sent by controllers
C. The hardware eventually failed
D. The mission was scheduled to end on this date

What is a Trojan asteroid?

A. One that shares an orbit with a planet or moon
B. One that orbits beyond Pluto
C. One that crosses orbits with the Earth
D. One that orbits closer to the Sun than Mercury

November 12th

Comet 67P/Churyumov–Gerasimenko was the destination for the first probe to land on a comet in 2014. Credit: DLR.

Which probe became the first to land intact on the surface of a comet on this date in 2014?

A. *Deep Impact*
B. *Philae*
C. *Stardust*
D. *Suisei*

How long does it take Neptune to complete one orbit of the Sun?

A. 164.8 years
B. 167.9 years
C. 170.3 years
D. 172.1 years

November 13th

The Tartarus Doras region of Pluto has terrain that's reminiscent of snakeskin. Credit: NASA.

Which probe became the first to orbit Mars on this date in 1971?

A. *Mariner 9*
B. *Mars Pathfinder*
C. *Mars 2*
D. *Viking 1*

What type of ice covers more than 98% of Pluto's surface?

A. Carbon-dioxide ice
B. Methane ice
C. Nitrogen ice
D. Water ice

November 14th

Vesta was the fourth asteroid to be discovered. Credit: NASA/JPL/MPS/DLR/IDA/Björn Jónsson.

Which of the following asteroids regularly appears brightest in our night sky?

A. Ceres
B. Pallas
C. Juno
D. Vesta

Which astronomical phenomenon did the ancient Greek astronomer Aristotle attribute to disturbances in the atmosphere?

A. Aurorae
B. Comets
C. Total lunar eclipses
D. Total solar eclipses

November 15th

The constellation Cassiopeia has a distinctive crooked W shape in the night sky. Credit: NASA/JPL-Caltech/DSS

Who does the constellation of Cassiopeia represent n Greek mythology?

A. A goddess and wife of Zeus
B. A hunter and sister of Orion
C. A servant girl and lover of Hercules
D. A queen and mother of Andromeda

Roughly how many stars can you see from a suburban location?

A. 300
B. 350
C. 400
D. 450

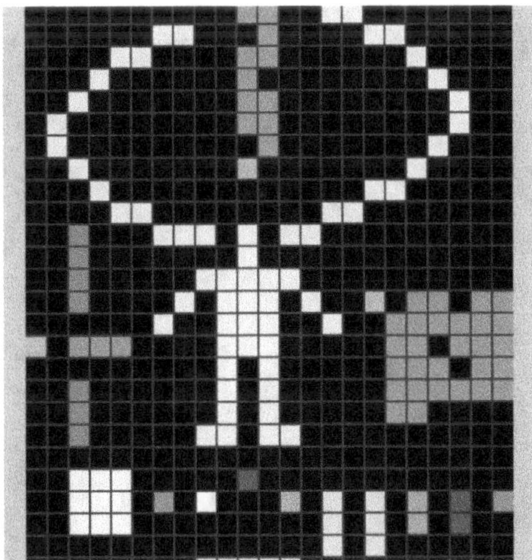

Part of the Arecibo Message with the Sun and planets at the bottom. The Earth's position is indicated by the raised square below the figure's feet. Credit: Arne Nordmann.

The Arecibo Message, consisting of data detailing DNA and our solar system, is broadcast on this date in 1974. What was the intended target?

A. Alpha Centauri
B. The Keystone globular star cluster
C. The Pleiades open star cluster
D. Vega

Which planet has the hottest surface temperature?

A. Venus
B. Mercury
C. Jupiter
D. Mars

November 17th

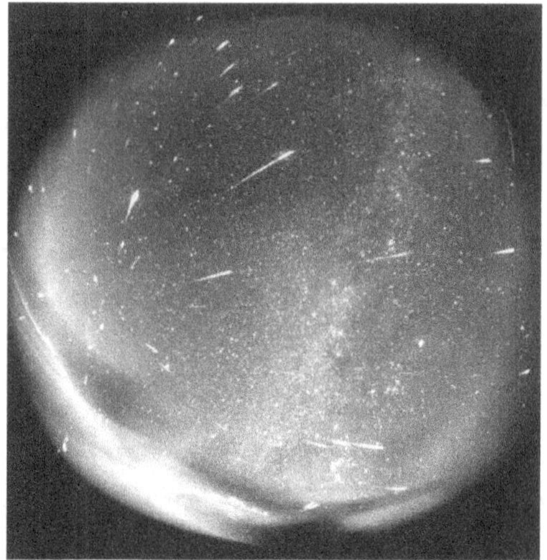

This fish-eye view of the whole sky shows a number of bright fireballs during the Leonid meteor shower of 1998. Credit: Comenius University.

In what year did the Leonid meteor shower produce a storm of shooting stars at a rate of 60,000 an hour?

A. 1956
B. 1961
C. 1966
D. 1971

Approximately how frequently does the Leonid meteor shower produce a strong storm?

A. Every 30 years
B. Every 33 years
C. Every 36 years
D. Every 39 years

November 18th

The Atacama Pathfinder Experiment telescope looks to the sky in northern Chile. Credit: ESO/B. Tafreshi

At an altitude of 5,640 meters (18,500 feet) which location has the highest observatory in the world?

A. The Andes, Bolivia
B. Atacama Desert, Chile
C. Salta, Argentina
D. The Tibet Autonomous Region, China

What name did William Herschel originally suggest for Uranus, the planet he discovered in 1781?

A. George's Star
B. Herschel
C. Minerva
D. Poseidon

November 19th

This artist's impression shows one of the Voyager *probes as it traverses interstellar space. Credit: NASA/JPL.*

What was the first man-made object to enter interstellar space?

A. *Pioneer 10*
B. *Pioneer 11*
C. *Voyager 1*
D. *Voyager 2*

For how long do the poles of Uranus spend in darkness as it orbits the Sun?

A. 22 years
B. 28 years
C. 34 years
D. 42 years

November 20th

Callisto is the most heavily cratered moon of Jupiter and is the third largest moon in the solar system. This image was taken by the Voyager 2 *probe in July, 1979. Credit: NASA*

Which moon of Jupiter's is the largest?

A. Io
B. Europa
C. Ganymede
D. Callisto

What gas makes up 80% of Neptune's atmosphere?

A. Helium
B. Hydrogen
C. Methane
D. Nitrogen

November 21st

Sirius, the brightest star in the sky, can be found by following the three stars of Orion's belt toward the south. Credit: Akira Fuji.

How many exoplanets are known to exist?

A. Less than 1,500
B. Between 1,500 and 2,500
C. Between 2,500 and 3,500
D. More than 3,500

Which star's name means "the follower"?

A. Aldebaran
B. Betelgeuse
C. Procyon
D. Sirius

November 22nd

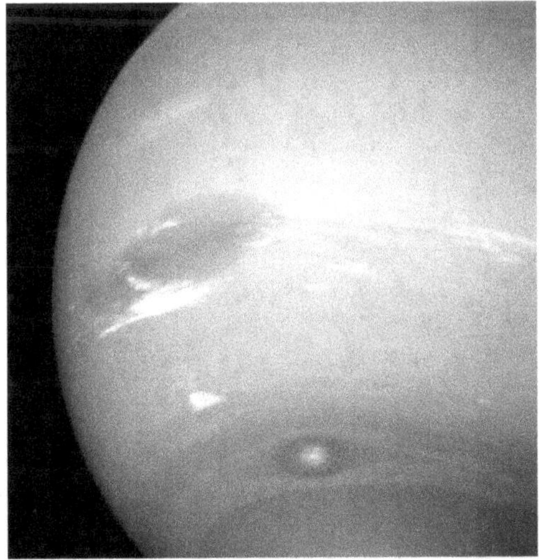

In this image from Voyager 2, *a white circular cloud can clearly be seen in the southern hemisphere of the planet. Credit: NASA*

What important discovery did NASA scientists announce on this date in 2016?

A. The first potentially habitable exoplanet
B. Fossilized microbial life in a Martian meteorite
C. A black monolith on the surface of the Moon
D. A frozen lake beneath the surface of Mars

What nickname was given to a white cloud imaged by *Voyager 2* during its approach to the planet Neptune?

A. The Ghost
B. The Marshmallow
C. Mr Stay-Puft
D. The Scooter

November 23rd

This artist's depiction shows ULAS J1120+0641, a quasar that formed just 770 million years after the Big Bang. Credit: ESO/M. Kornmesser

What significant discovery did Edwin Hubble announce on this date in 1924?

A. That the universe began with the Big Bang
B. That the universe was expanding
C. That the Andromeda Nebula is actually a galaxy
D. That nothing travels faster than light

Which moon of Neptune has an atmosphere that's primarily comprised of nitrogen?

A. Larissa
B. Triton
C. Neried
D. Thalassa

November 24th

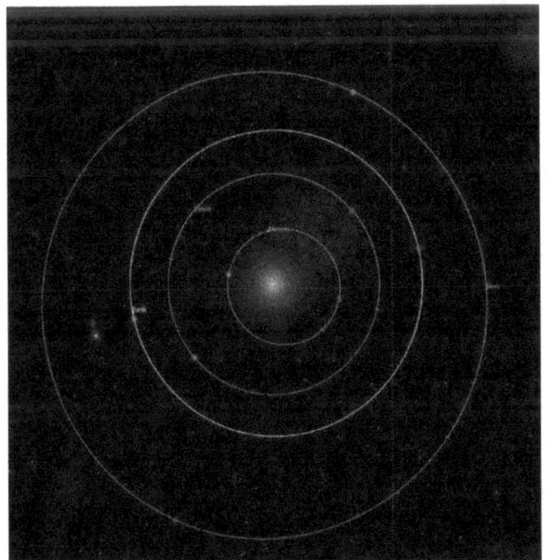

This image shows the orbits of the four innermost planets to scale. Innermost is Mercury, then Venus, Earth and finally Mars. Credit: Danial79 (from Wikimedia.)

Which planet has the most elliptical orbit?

A. Mercury
B. Earth
C. Jupiter
D. Uranus

Which of the following worlds may be visible with the naked eye under ideal conditions?

A. Chiron (asteroid)
B. Uranus
C. Neptune
D. Pluto

November 25th

This unusual view of the Moon's north pole was taken by the Galileo *space probe while en-route to Jupiter on December 7th, 1992. Credit: NASA/JPL.*

Which moon is the densest in the solar system?

A. Io (Jupiter)
B. The Moon (Earth)
C. Titan (Saturn)
D. Charon (Pluto)

Which of the following is *not* visible to the naked eye?

A. Andromeda Galaxy
B. The Hyades
C. The Horsehead Nebula
D. The Pleiades

November 26th

The astronomer Copernicus has a prominent crater on the Moon named for him. Credit: NASA.

What is Nicolaus Copernicus most famous for?

A. Discovering the four largest moons of Jupiter
B. The invention of the telescope
C. Proposing that the Earth was flat
D. Proposing that the Earth orbited the Sun

When were the first recorded observations of Mercury made?

A. 14th century B.C.
B. 5th century B.C.
C. 2nd century A.D.
D. 7th century A.D.

November 27th

Saturn could fit roughly 763 Earths within it. Credit: NASA.

What's the average temperature on the surface of Mars?

A. -11° C (12° F)
B. -34° C (-29° F)
C. -63° C (-81° F)
D. -72° C (-98° F)

Which planet has a radius roughly nine times that of the Earth?

A. Jupiter
B. Saturn
C. Uranus
D. Neptune

November 28th

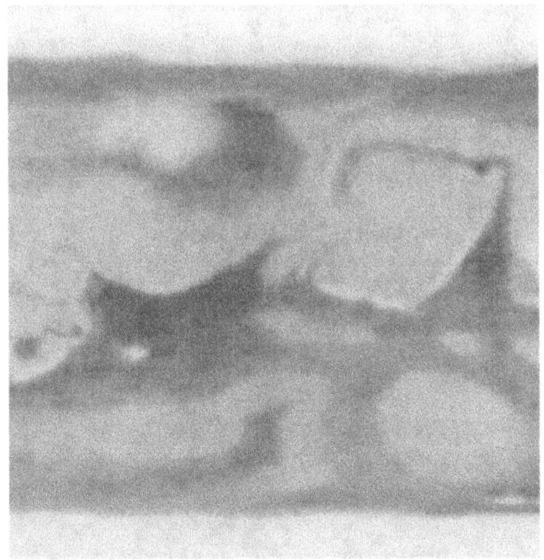

This map of Mars was produced by the astronomer Nathaniel E. Green in 1877. Credit: Public Domain.

Who was the first astronomer to produce a map of Mars in 1659?

A. Giovanni Cassini
B. Galileo Galilei
C. Christiaan Huygens
D. Johannes Kepler

Which planet's rings are the least stable?

A. Jupiter
B. Saturn
C. Uranus
D. Neptune

November 29th

This image mosiac shows the north pole of Mercury. Credit: NASA/Johns Hopkins University Applied Physics Laboratory/Carnegie Institution of Washington.

What discovery is announced by scientists on this date in 2012?

A. Fossilized microbial life on Mars
B. Faint traces of oxygen in the Moon's atmosphere
C. Organic molecules in the atmosphere of Titan
D. Water ice at the north pole of Mercury

Which of the following stars is not a member of the Pleiades open star cluster?

A. Alcor
B. Celaeno
C. Pleione
D. Sterope

November 30th

The four innermost planets orbit closer to the Sun than Jupiter (upper left.) Credit: NASA/JPL.

Which of the following is the largest of the four innermost planets?

A. Mercury
B. Venus
C. Earth
D. Mars

Uranus's moons Titania and Oberon are named after characters from which Shakespearean play?

A. Macbeth
B. A Midsummer Night's Dream
C. Romeo and Juliet
D. The Tempest

December

December 1st

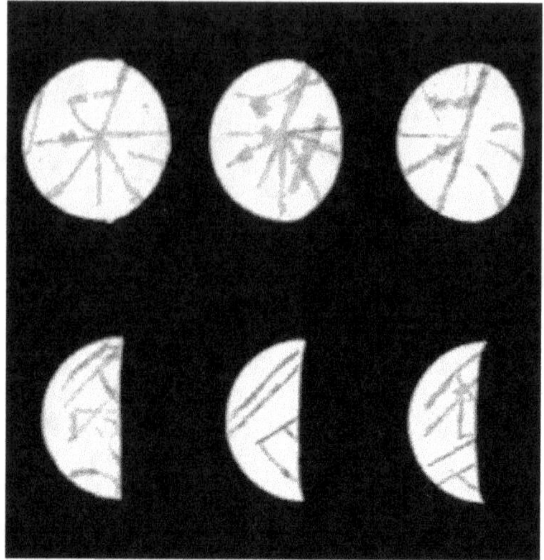

These sketches of Venus through a telescope were first published in 1909. Credit: Public domain.

Christiaan Huygens makes the first reasonably accurate estimate of the rotation period of which planet on this date in 1659?

A. Jupiter
B. Venus
C. Mercury
D. Mars

What's the average distance between Venus and the Sun? (1 AU = the distance between the Earth and the Sun)

A. 0.64 AU
B. 0.72 AU
C. 0.81 AU
D. 0.93 AU

December 2nd

The surface of Mars, as seen from the landing site of Viking 1. *Credit: NASA/JPL.*

Which probe was the first to land on Mars in 1971?

A. *Mariner 9*
B. *Mars 3*
C. *Pathfinder*
D. *Viking 1*

In which month is the Earth closest the Sun?

A. January
B. April
C. July
D. October

December 3rd

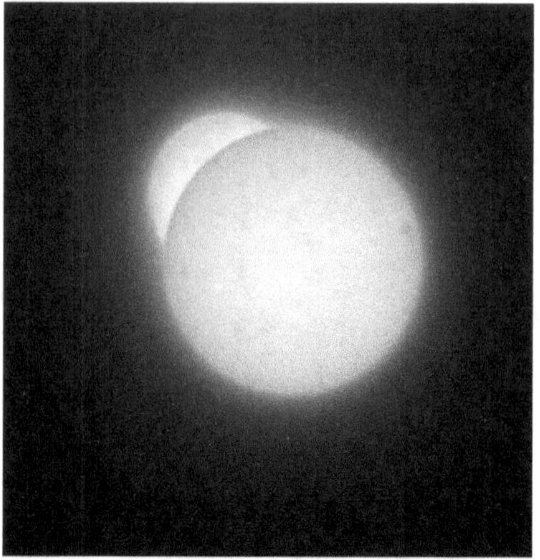

This artist's impression depicts a star system, similar to Algol. When one star passes behind the other, the star appears to temporarily dim. Credit: ESO/L. Calçada.

In what year did the first observation of Venus transiting the Sun occur?

A. 1639
B. 1644
C. 1649
D. 1654

The variable star Algol dims and brightens over what of time?

A. 1 day, 15 hours, 23 minutes
B. 1 day, 21 hours, 15 minutes
C. 2 days, 8 hours, 31 minutes
D. 2 days, 20 hours, 49 minutes

December 4th

In this artist's impression, what space probe is depicted passing the planet Jupiter? Credit: NASA/Rick Guidice

Observers at which observatory were said to have observed a shaft of light emanating from Mars for 70 minutes in December 1900?

A. Griffith Observatory at Los Angeles, California
B. Lowell Observatory at Flagstaff, Arizona
C. Royal Observatory at Greenwich, England
D. Yerkes Observatory at Williams Bay, Wisconsin

Which space probe became the first to pass the planet Jupiter on this date in 1973?

A. Juno
B. Pioneer 9
C. Pioneer 10
D. Voyager 1

December 5th

Voyager 2 *didn't pass very close to Uranus's moon Umbriel when it flew by the planet in January 1986. Credit: NASA/JPL.*

Which planet has moons named after characters from Shakespearean plays?

A. Jupiter
B. Saturn
C. Uranus
D. Neptune

What chemical gives some meteors an orange-yellow color?

A. Carbon
B. Magnesium
C. Potassium
D. Sodium

The full Moon has been a familiar sight for billions of people across thousands of years. Credit: Author.

Which astronomer first proposed the name "asteroid" for the newly discovered objects found between the orbits of Mars and Jupiter?

A. Edmond Halley
B. William Herschel
C. Isaac Newton
D. Giuseppe Piazzi

At what time of day does the full Moon rise?

A. At sunrise
B. At midday
C. At sunset
D. At midnight

December 7th

Artist Ken Hodges depicts Galileo's *atmospheric probe as it plunges into Jupiter's atmosphere. Credit: NASA/JPL.*

What causes the variable star Algol to dim and then brighten again?

A. A black hole passing in front of a bright star
B. An orbiting planet eclipsing the star
C. The star expanding and contracting
D. A bright star eclipsed by a fainter companion

For how long did the atmospheric probe of the *Galileo* space craft last when it entered the atmosphere of Jupiter on this date in 1995?

A. 57 minutes
B. 1 hour, 10 minutes
C. 1 hour, 48 minutes
D. 2 hours, 3 minutes

Jupiter's moon Ganymede is the largest and most massive moon in the solar system. Credit: NOAA.

In what year did the *Galileo* space probe become the first to orbit Jupiter?

A. 1994
B. 1995
C. 1996
D. 1997

Which of Jupiter's four Galilean satellites is the only one *not* suspected of having a subsurface ocean?

A. Callisto
B. Europa
C. Ganymede
D. Io

December 9th

The Pleiades star cluster can easily be seen with the naked eye in the autumn/winter constellation of Taurus. Credit: Anttler (from Wikimedia.)

Which poem by Tennyson describes the Pleiades open star cluster as "fireflies?"

A. *Locksley Hall*
B. *Miriana*
C. *Tithonus*
D. *Ulysses*

What's the total surface area of the Earth?

A. 350 million square kilometers
B. 420 million square kilometers
C. 510 million square kilometers
D. 625 million square kilometers

The major moons of the solar system, shown to scale, with the Earth in the top left corner for comparison. Credit: NASA

How many moons in the solar system are known to be larger than our own Moon?

A. Two
B. Four
C. Six
D. Eight

How bright is the sunlight at Uranus compared to the sunlight on Earth?

A. 1/50th
B. 1/100th
C. 1/200th
D. 1/400th

December 11th

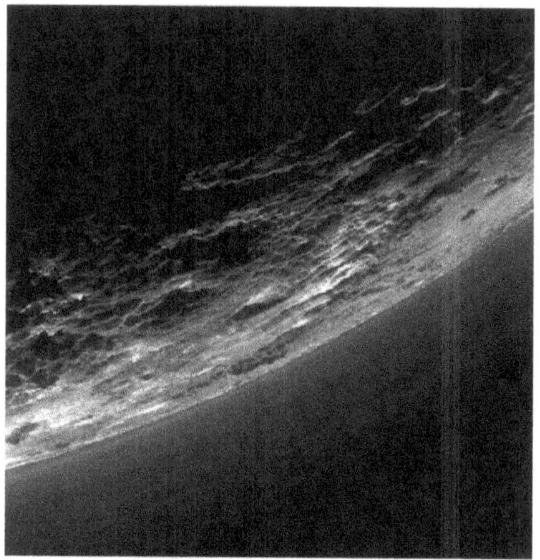

This image from the New Horizons *space probe clearly shows the thin haze of Pluto's atmosphere. Credit: NASA/Johns Hopkins University Applied Physics Laboratory/Southwest Research Institute/AAAS/Science*

What role did the god Pluto have in Greek mythology?

A. The bringer of death
B. The god of money and material possessions
C. God of the ocean
D. God of the underworld

Which gas makes up 78% of the Earth's atmosphere?

A. Carbon Dioxide
B. Helium
C. Nitrogen
D. Oxygen

Orion is clearly visible to the right in this image with Sirius shining brilliantly near the bottom. Procyon, the brightest star in Canis Minor, the Little Dog, appears to the upper left. Credit: Akira Fuji/Hubble European Space Agency.

What is the Earth's orbital velocity?

A. 21.2 km/sec (13.2 miles/sec)
B. 29.8 km/sec (18.5 miles/sec)
C. 38.1 km/sec (23.7 miles/sec)
D. 45.2 km/sec (28.1 miles/sec)

Which star's name means "glowing" or "scorching"?

A. Antares
B. Procyon
C. Sirius
D. Vega

December 13th

The inner planets, from top left to bottom right (and in order of distance from the Sun): Mercury, Venus, Earth and Mars. Credit: NASA

Which planet is almost the same size as the Earth?

A. Venus
B. Mars
C. Mercury
D. Neptune

The Earth's orbital speed is 29.78 km/sec. How long does it take the Earth to move through space a distance equal to its diameter?

A. 7 minutes
B. 13 minutes
C. 19 minutes
D. 25 minutes

Under ideal conditions, the Geminids can produce a spectacular show. Credit: John Flannery

On this date in 1962 the space probe *Mariner 2* became the first to successfully fly by another planet and send back data. Which planet did it pass?

A. Mercury
B. Venus
C. Mars
D. Jupiter

The Geminid meteor shower peaks around this date. How many meteors can you see hourly under ideal conditions?

A. 60
B. 80
C. 100
D. 120

December 15th

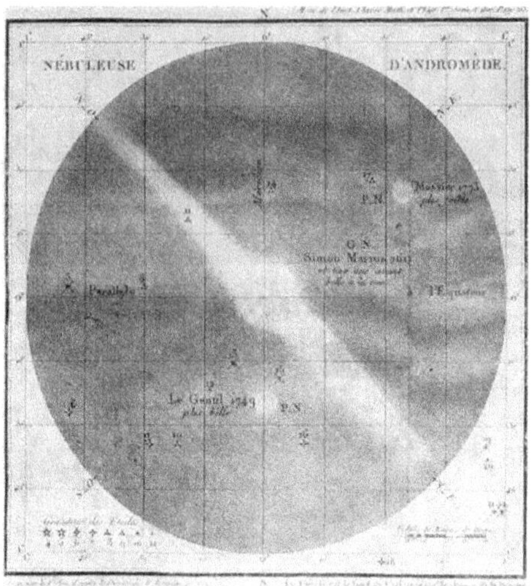

This sketch of the Andromeda Galaxy by the great French astronomer Charles Messier dates back to 1807. Credit: Public domain.

Who was the first to record a telescopic observation of the Andromeda Galaxy on this date in 1612?

A. Giovanni Cassini
B. Galileo Galilei
C. Christiaan Huygens
D. Simon Marius

Which probe was the first to successfully land on the surface of Venus and send back data on this date in 1970?

A. Venera 7
B. Venera 8
C. Venera 9
D. Venera 10

December 16th

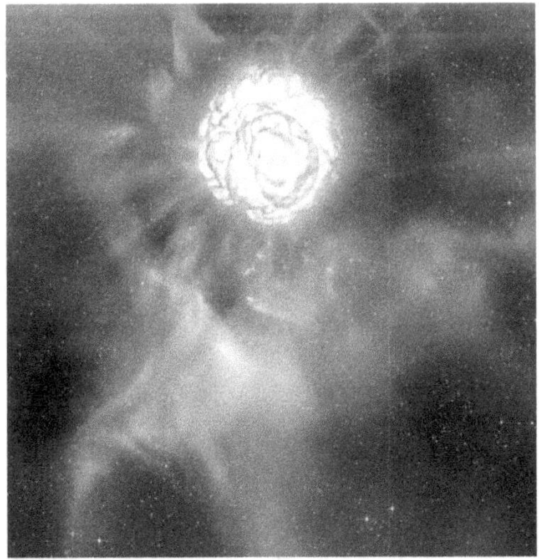

This artist's impression depicts the vast plume of gas observed erupting from the supergiant star Betelgeuse. Credit: ESO/L. Calçada

Which planet is the only one whose name is directly derived from that of a Greek god?

A. Mars
B. Jupiter
C. Saturn
D. Uranus

Which of the following stars *cannot* be found in the constellation Orion?

A. Alnitak
B. Betelgeuse
C. El Nath
D. Rigel

December 17th

Saturn's moon Mimas (lower left) appears against the backdrop of the planet's atmosphere and the shadows cast by Saturn's rings. Credit: NASA/JPL/Space Science Institute

On this date in 1900, the Guzman Prize offered 100,000 francs to the first person to do what?

A. Communicate with extra-terrestrials
B. Discover the ninth planet
C. Fly to the Moon and back
D. Prove the existence of life on Mars

Which moon of Saturn has one bright and one dark hemisphere?

A. Enceladus
B. Iapetus
C. Mimas
D. Titan

December 18th

Orion is one of the brightest and most recognizable constellations in the December night sky. Credit: TheStarmon (from Wikimedia)

What's the only planet in the solar system with just one moon?

A. Mercury
B. Earth
C. Mars
D. Neptune

Which two stars mark the shoulders of Orion, the Hunter?

A. Bellatrix and Rigel
B. Bellatrix and Saiph
C. Betelgeuse and Bellatrix
D. Betelgeuse and Rigel

December 19th

An artist's impression of the Tau Ceti star system. Credit: J. Pinfield for the RoPACS network at the University of Hertfordshire, 2012

Scientists announce the suspected discovery of planets orbiting the nearby star Tau Ceti on this date in 2012. How many planets are thought to orbit the star?

A. Two
B. Three
C. Four
D. Five

How old are the rings of Saturn thought to be?

A. Hundreds of thousands of years old
B. About a million years old
C. Hundreds of millions of years old
D. About a billion years old

December 20th

Galileo Galilei discovered Jupiter's four largest moons in January, 1610. This painting dates back to 1636. Credit: Justus Suttermans

Why was the Italian astronomer Galileo Galilei under house arrest from 1634 until his death in 1642?

A. He believed the Earth went around the Sun
B. He believed the Earth was a sphere
C. He believed in life on other planets
D. He no longer believed in God

Which planet is the only one to have moons that are theoretically visible to the naked eye?

A. Mars
B. Jupiter
C. Saturn
D. Uranus

December 21st

The Mariner 4 *space probe was commemorated in this stamp from Hungary. Credit: Alf van Beem.*

The winter solstice occurs around this date in the northern hemisphere. What's the official name for this event?

A. Autumnal equinox
B. Estival solstice
C. Hibernal solstice
D. Vernal equinox

The *Mariner 4* mission comes to an end today in 1967. The first successful probe to Mars and to take close-up images of another planet, how many photos did it send back?

A. 21
B. 36
C. 51
D. 66

December 22nd

This image depicts the total solar eclipse of March 29th, 2006 and was taken near Tokat, Turkey. Credit: Kurt Kulac.

On this date in 1870 the French astronomer Jules Janssen escaped the besieged city of Paris to observe a total solar eclipse. How did he make his escape?

A. By car
B. By foot
C. By horse
D. By hot air balloon

Max Wolf discovered the asteroid Brucia on this date in 1891. What made this asteroid unique?

A. It had the shortest orbital period at the time
B. It was the first found inside the orbit of Mars
C. It was first to be discovered photographically
D. It was the faintest asteroid known at the time

December 23rd

The Cassini *probe (top) flies over the surface of Enceladus in this artist's impression. Credit: NASA/JPL/David Seal.*

Giovanni Cassini discovered the third moon of Saturn on this date in 1672. Which moon was discovered?

A. Enceladus
B. Rhea
C. Tethys
D. Titan

On this date in 1690 John Flamsteed unknowingly made the first recorded observation of which undiscovered object?

A. Ceres (dwarf planet)
B. Neptune
C. Pluto
D. Uranus

December 24th

Artists' impression of the surface of Pluto with its largest moon Charon hanging over the horizon. Credit: ESO/L. Calçada

What's the mean temperature on the surface of Pluto?

A. -157° C (-251° F)
B. -183° C (-297° F)
C. -229° C (-380° F)
D. -256° C (-429° F)

If the Sun were hollow, roughly how many Earths could fit within it?

A. 13 thousand
B. 103 thousand
C. 1.3 million
D. 10.3 million

December 25th

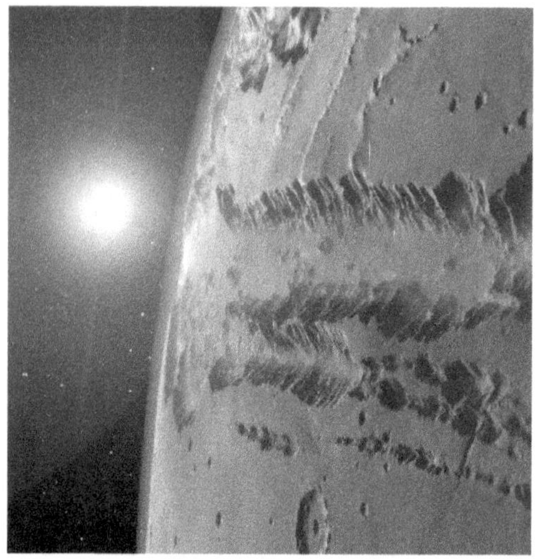

An artist's depiction looking down upon the largest valley on Mars from orbit. Credit: Kevin Gill

Which famous astronomer was born on this day in 1642?

A. Giovanni Cassini
B. Edmond Halley
C. Christiaan Huygens
D. Isaac Newton

Which valley on Mars is 4,000 km (2,485 miles) long, 7km deep and extends across one fifth the surface of the planet?

A. Ares Vallis
B. Valles Marineris
C. Tiu Vallis
D. Mangala Vallis

December 26th

The open star cluster Messier 35 can be found at the feet of Gemini, the Twins. Credit: Oliver Stein

Which of the following is not a winter constellation?

A. Virgo
B. Orion
C. Canis Major
D. Gemini

How much brighter does the Sun appear in our daytime sky compared to Sirius, the brightest star in our nighttime sky?

A. 13 hundred times
B. 13 thousand times
C. 13 million times
D. 13 billion times

December 27th

On June 30th, 2015, Venus (bottom) and Jupiter (top) appeared close together in the evening twilight. Credit: Pmau (from Wikimedia).

What color does Venus appear in our sky?

A. Red
B. White
C. Blue
D. Yellow

If Neptune were hollow, how many Earths could fit inside it?

A. 38 Earths
B. 42 Earths
C. 57 Earths
D. 61 Earths

December 28th

This artist's impression shows Haumea, an ellipsoid dwarf planet and its two small moons, Hi'aka and Namaka. Credit: José Antonio Peñas/SINC

Mike Brown and his team discover the dwarf planet Haumea on this date in 2004. What nickname was given to it by the team?

A. Blitzen
B. Donner
C. Rudolph
D. Santa

What color does Saturn appear in our sky?

A. Orange
B. Yellow
C. White
D. Red

December 29th

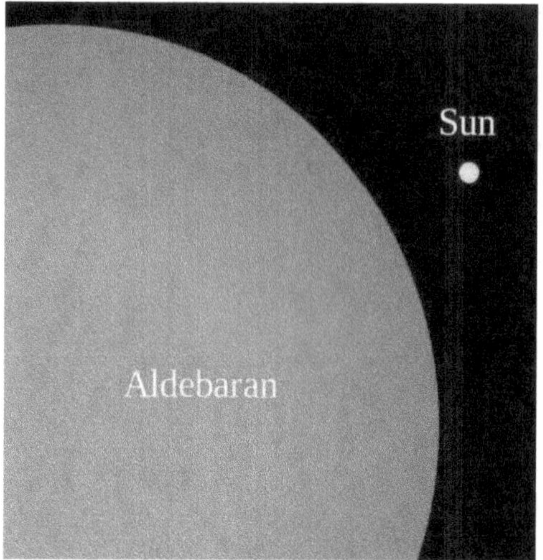

Aldebaran is an orange giant star that lies about 65 lights years away and is thought to be roughly 44 times larger than the Sun. Credit: Mysid (from Wikimedia)

Which planet is the densest?

A. Mercury
B. Earth
C. Mars
D. Jupiter

Which star marks the red eye of Taurus the Bull?

A. Aldebaran
B. Castor
C. Antares
D. Sirius

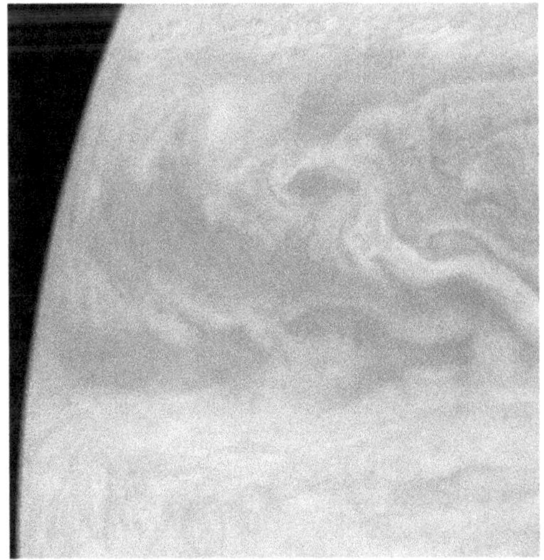

The Cassini *space probe took this close-up image of Jupiter's cloud-tops just one day before its flyby of the planet. Credit: NASA/JPL/Space Science Institute*

Which planet did the *Cassini-Huygens* probe pass on this date in 2000?

A. Earth
B. Jupiter
C. Mars
D. Venus

In 2009 it has announced that the star Betelgeuse has shrunk since 1993 – but by how much?

A. 6%
B. 9%
C. 12%
D. 15%

December 31st

This false color image of Iapetus was taken on September 10th, 2007, at a distance of approximately 73,000 km (45,300 miles.) Credit: NASA/JPL/Space Science Institute

What feature on Saturn's moon Iapetus did the space probe *Cassini* discover on this date in 2004?

A. The largest known crater on a moon
B. An equatorial ridge
C. A subsurface ocean
D. Frozen water at the north pole

Which moon has river networks on its surface?

A. Europa (Jupiter)
B. Miranda (Uranus)
C. Titan (Saturn)
D. Triton (Neptune)

Answers

1st

January	D B	**July**	C C
February	B A	**August**	D A
March	C B	**September**	D B
April	D A	**October**	B A
May	D C	**November**	B A
June	A A	**December**	D B

2nd

January	A A	**July**	A B
February	B D	**August**	B A
March	A D	**September**	A D
April	B A	**October**	D D
May	C A	**November**	D C
June	B D	**December**	B A

3rd

January	C A	**July**	B A
February	A C	**August**	D D
March	B D	**September**	D B
April	D A	**October**	B A
May	D B	**November**	B D
June	D B	**December**	A D

4th

January	D A	**July**	D C
February	B D	**August**	B D
March	D C	**September**	C C
April	C A	**October**	B C
May	A A	**November**	D A
June	B C	**December**	B C

5th

January	B A	**July**	C C
February	C D	**August**	A A
March	B A	**September**	A D
April	C D	**October**	A D
May	D C	**November**	A C
June	A B	**December**	C D

6th

January	D B	**July**	A B
February	D A	**August**	A B
March	C D	**September**	D B
April	B B	**October**	B A
May	B D	**November**	C A
June	D D	**December**	B C

7th

January	A B	**July**	D B
February	D B	**August**	C A
March	A D	**September**	D B
April	A C	**October**	D B
May	B A	**November**	A D
June	A C	**December**	D A

8th

January	D D	**July**	B C
February	C B	**August**	A C
March	A C	**September**	C A
April	B D	**October**	B A
May	C D	**November**	C C
June	B A	**December**	B D

9th

January	D A	**July**	B D
February	C D	**August**	D B
March	C A	**September**	A A
April	B A	**October**	A C
May	B C	**November**	D D
June	C D	**December**	A C

10th

January	C B	**July**	C B
February	B A	**August**	D B
March	A A	**September**	B A
April	C D	**October**	B D
May	A A	**November**	C B
June	B D	**December**	B D

11th

January	D A	**July**	B D
February	C B	**August**	C A
March	B C	**September**	B C
April	A B	**October**	A B
May	C C	**November**	B A
June	D A	**December**	D C

12th

January	C A	**July**	D A
February	D A	**August**	C B
March	A D	**September**	C A
April	C A	**October**	A C
May	C D	**November**	B A
June	C A	**December**	B C

13th

January	B C	**July**	B A
February	C B	**August**	D C
March	B D	**September**	A B
April	C D	**October**	D D
May	A D	**November**	A C
June	A C	**December**	A A

14th

January	C A	**July**	A C
February	B C	**August**	D A
March	D D	**September**	D C
April	D A	**October**	C B
May	A B	**November**	D B
June	C D	**December**	B D

15th

January	B D	**July**	B C
February	A C	**August**	C A
March	D A	**September**	A D
April	B C	**October**	B D
May	B A	**November**	D D
June	D B	**December**	D A

16th

January	D C	**July**	D A
February	C D	**August**	B B
March	C A	**September**	B C
April	D B	**October**	A B
May	D C	**November**	B A
June	A B	**December**	D C

17th

January	A B	**July**	D C
February	C B	**August**	C B
March	A D	**September**	C D
April	B C	**October**	C D
May	C A	**November**	C B
June	D C	**December**	A B

18th

January	D C	**July**	C B
February	B A	**August**	A D
March	D A	**September**	C A
April	A A	**October**	C C
May	D B	**November**	B A
June	C C	**December**	B C

19th

January	D A	**July**	A A
February	D C	**August**	B C
March	B C	**September**	B D
April	D B	**October**	C B
May	A D	**November**	C D
June	B D	**December**	D C

20th

January	D A	**July**	D C
February	A C	**August**	D D
March	A D	**September**	A B
April	C A	**October**	D A
May	B A	**November**	C B
June	C D	**December**	A B

21st

January	C A	**July**	B D
February	D C	**August**	C B
March	A D	**September**	B C
April	A D	**October**	B C
May	D B	**November**	D A
June	B B	**December**	C A

22nd

January	B A	**July**	A C
February	C B	**August**	D C
March	A C	**September**	D A
April	B D	**October**	A D
May	D B	**November**	D D
June	C A	**December**	D C

23rd

January	A D	**July**	B C
February	C D	**August**	C B
March	B C	**September**	A A
April	B B	**October**	A C
May	A D	**November**	C B
June	D B	**December**	B D

24th

January	B A	**July**	B A
February	D D	**August**	D C
March	C A	**September**	D C
April	A D	**October**	C B
May	A C	**November**	A B
June	B D	**December**	C C

25th

January	A D	**July**	C C
February	C D	**August**	C B
March	D D	**September**	B C
April	A C	**October**	B A
May	C A	**November**	A C
June	A C	**December**	D B

26th

January	B C	**July**	B A
February	D A	**August**	C C
March	B A	**September**	D A
April	C B	**October**	C D
May	C B	**November**	D A
June	B A	**December**	A D

27th

January	B C	**July**	C A
February	B D	**August**	B D
March	A C	**September**	B C
April	B A	**October**	C A
May	B A	**November**	C B
June	A C	**December**	B C

28th

January	B D	**July**	B D
February	C D	**August**	B A
March	C B	**September**	C A
April	C C	**October**	A D
May	D B	**November**	C D
June	B D	**December**	D B

29th

January	B C	**July**	C D
February	A B	**August**	B C
March	A C	**September**	D C
April	A B	**October**	C D
May	C D	**November**	D A
June	C D	**December**	B A

30th

January	C A	**July**	C B
		August	C B
March	B A	**September**	D A
April	A C	**October**	B C
May	B D	**November**	C B
June	A B	**December**	B D

31st

January	C B	**July**	A D
		August	B D
March	C A		
		October	D B
May	B D		
		December	B C

Also by the Author

Signposts to the Stars

Aimed at absolute beginners, this book will help you to locate and learn the constellations using the brightest stars of Ursa Major and Orion as signposts.

More than that, the book also details:

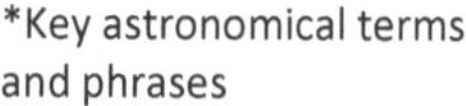

*Key astronomical terms and phrases
*The brightest stars and constellations for each season
*The myths and legends of the stars
*Fascinating stars, star clusters, nebulae and galaxies, many of which can be seen with just your eyes or binoculars
*An introduction to the planets, comets and meteor showers

If you've ever stopped and stared at the stars but didn't know where to begin, these signposts will get you started on your journey!

Easy Things to See With a Small Telescope

Specifically written with the beginner in mind, this book highlights over sixty objects easily found and observed in the night sky.

Objects such as:
* Stunning multiple stars
* Star clusters
* Nebulae
* And the Andromeda Galaxy!

Each object has its own page which includes a map, a view of the area through your finderscope and a depiction of the object through the eyepiece.

There's also a realistic description of every object based upon the author's own notes written over years of observations.

Additionally, there are useful tips and tricks designed to make your start in astronomy easier and pages to record your observations.

If you're new to astronomy and own a small telescope, this book is an invaluable introduction to the night sky.

The Easy Guide to the Night Sky

Written for the amateur astronomer who wants to discover more in the night sky, this book explores the constellations and reveals many of the highlights visible with just your eyes or binoculars.

The highlights include:

* The myths and legends associated with the stars
* Bright stars and multiple stars
* Star clusters
* Nebulae
* Galaxies

Each constellation has its own star chart and almost all are accompanied by graphics depicting the highlights and binocular views of the best objects.

Whether you're new to astronomy or are an experienced stargazer simply looking to learn more about the constellations, this book is an invaluable guide to the night sky and the stars to be found there.

The Deep Sky Observer's Guide

The Deep Sky Observer's Guide offers you the night sky at your fingertips. As an amateur astronomer, you want to know what's up tonight and you don't always have the time to plan ahead. The Deep Sky Observer's Guide can solve this problem in a conveniently sized paperback that easily fits in your back pocket. Take it outside and let the guide suggest any one of over 1,300 deep sky objects, all visible with a small telescope and many accessible via binoculars.

* Multiple stars with 2" or more of separation
* Open clusters up to magnitude 9
* Nebulae up to magnitude 10
* Globular clusters up to magnitude 10
* Planetary nebulae up to magnitude 12
* Galaxies up to magnitude 12
* Includes lists of deep sky objects for the entire sky with R.A. and declination for each and accompanying images for many

Whether you use a GoTo or prefer to star hop, no matter where you live in the world and no matter what time of year or night, the Deep Sky Observer's Guide is the indispensable companion for every adventure among the stars.

The Wonder of It All

From our home here on Earth, past the Sun, Moon and planets, this is a journey out to the stars and beyond.

THE WONDER OF IT ALL

RICHARD J. BARTLETT

A journey of discovery that shows us the beauty and wonder of the cosmos and our special and unique place within it.

Written by an amateur astronomer with a life-long love of the stars, The Wonder of It All will open your child's eyes to the universe and includes notes for parents to help develop an interest in astronomy.

The Amateur Astronomer's Notebook

The Amateur Astronomer's Notebook is the perfect way to log your observations of the Moon, stars, planets and deep sky objects.

With an additional appendix with hundreds of suggested deep sky objects, this 8.5" by 11" notebook allows you to record everything you need for 150 observing sessions under the stars:

*Date
*Time
*Lunar Phase
*Limiting Magnitude
*Transparency
*Seeing
*Equipment
*Eyepieces
*Additional Notes
*Pre-drawn circles to sketch your observations
*Plenty of room to record your notes and impressions

Whether you're an experienced astronomer or just beginning to discover the universe around us, you'll find the notebook to be an invaluable tool and record of your exploration of the cosmos.

The Astronomical Almanac

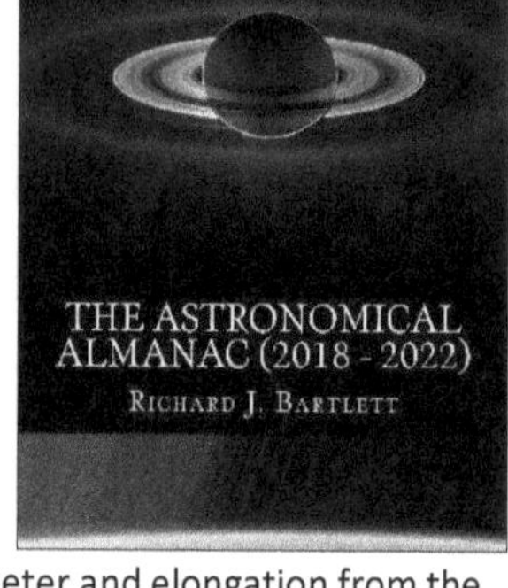

The guide provides almost daily data and information on the Moon and planets and lists details of hundreds of astronomical events.

* Graphical depictions of the Moon and planets throughout each month
* Data tables for the Moon and planets including position, magnitude, apparent diameter and elongation from the Sun.
* Conjunctions between the Moon, planets and bright stars (including angular separations for planetary conjunctions.)
* Lunar and Solar eclipses
* Annual summaries of when to observe the planets
* Annual summaries of notable close planetary conjunctions
* Peak dates for the major meteor showers with moon phase
* Inferior and superior conjunctions for Mercury and Venus
* Greatest eastern and western elongations for Mercury and Venus
* Oppositions and solar conjunctions for the outer planets
* Dates of the equinoxes and solstices

PLUS

* Twenty-four star charts with accompanying lists of suggested deep sky objects

www.ingramcontent.com/pod-product-compliance
Lightning Source LLC
Chambersburg PA
CBHW071247220526
45468CB00001B/28

* 9 7 8 1 9 8 3 8 2 7 4 3 3 *